Work quality management in the textile industry

Work quality management in the textile industry

B. Purushothama

WOODHEAD PUBLISHING INDIA PVT LTD

New Delhi

Published by Woodhead Publishing India Pvt. Ltd.
Woodhead Publishing India Pvt. Ltd.,
303, Vardaan House, 7/28, Ansari Road,
Daryaganj, New Delhi - 110002, India
www.woodheadpublishingindia.com

First published 2013, Woodhead Publishing India Pvt. Ltd.
© Woodhead Publishing India Pvt. Ltd., 2013
Reprint 2017, 2018
Reprinted 2020

Woodhead Publishing India Pvt. Ltd. ISBN: 978-93-80308-40-1
Woodhead Publishing India Pvt. Ltd. e-ISBN: 978-93-80308-68-5

Typeset by Mind Box Solutions, New Delhi
Printed and bound by Replika Press Pvt. Ltd.

Contents

Being the oldest among industries, the textile and garment industry has taken a significant role in the industrial revolution, development of work norms, development of number of statutory, legal and regulatory requirements, development of new management techniques, development of norms for industrial relations and so on. In spite of the industry being the oldest and has undergone various ups and downs, even today it is not in a position stabilize self and be a role model for other industries. The problems faced by the industry and the employees rather getting solved are getting increased. Developments of technology, automations, computer-aided techniques, etc., have helped the industry in getting the productivity and quality, but the same is not getting sustained. Customers are able to clearly precise the quality they require, and in the fashion world all the earlier so-called mistakes or poor quality are getting a different name as highly fashionable. The people do not prefer to work in textile or garment industry due to various reasons, and the managements are not trying to retain the people interested in working in the industry.

The textile and garment industry, which was once the backbone of advanced countries, has lost its base and has shifted to developing countries. India being the cradle of civilization and mother of textiles naturally has an edge, but still the industry is not doing well. The managements are blaming staff and workers and the employees are blaming management. The managements want the workers to give more efficiency but are overloading them with continuous working, unstable administration and are respecting them as a part of their industry.

All are trying to achieve quality and productivity by installing latest technology and paying huge salaries to the top persons in the organization, but are not addressing the basic requirements of clean administration, improving the quality of work, developing harmony among the staff and workers, and bringing a feeling of oneness among all. The people with power are carried away by the short-term plans as it looks lucrative and are not making any efforts to make the base stronger.

In this book "Work Quality Management in the Textile Industry", an attempt is made to explain the importance of maintaining "work quality", which can help the industry to achieve results in long term and can make it stable. Being a shop floor person working with various levels for past 43 years, I have tried to explain the concepts in the simplest possible terms, and examples given are actual; there is no fiction. I shall be happy if the readers take the concepts seriously and make efforts to come out from the present crises.

B. Purushothama
Consultant QMS and Textiles

1
What is work quality?

1.1   Introduction to work quality

Everyone in the society is aware of the term 'quality' and expects the services and products he is buying to be of the best quality. So, technology is in the continuous process of development to produce the best-quality products. Mainly, industries are categorised into two types: product manufacturing and service providing. Textile industry comes under the product manufacturing category.

Now, the big question is "what makes the quality?" When the same technology is adopted at different textile mills, do they get the same quality all the time? Why there is a difference in quality of the products of both the mills? Even after spending billions in buying latest machineries and best raw material and hiring the best people, manufacturing companies don't achieve the best quality. The correct answer to these questions is the difference in work quality of both the mills.

The highly motivated human resource of the company is more important to get the best quality. To drive the workers to achieve the quality should be emphasized more than anything else. When the management succeed in involving the people at work, the quality can be achieved even with old machines; of course, the productivity may be less. What additional work they do when work by heart? Do they adopt any different settings or speeds? No, it is the same machine, same setting, but when the work is done devotedly, the mistakes are not there. Only good quality is produced.

Producing a good quality product also depends on ensuring good raw material, adopting appropriate technology, tuning the machines appropriately and handling the machines appropriately. But how these are handled, preserved and maintained depends on the quality of work. Examples of poor work quality are the following: producing and stocking excess in advance,, reworking to correct the products to meet specifications, spending money on inspections to ensure that the product produced meets the requirements, keeping very narrow tolerance to avoid rejections, using the raw material of higher quality than required to avoid rejections, employing additional people or resources to ensure work is done on time, spending on advertisements and promotion disproportionately to the value of materials being sold, polluting the environment, not adhering to safety and legal regulations, paying less than what was supposed to paid, harassing the suppliers and employees and not providing a clean and healthy work environment. Quality of work is different from quality of product.

When a person does a work with interest and ensures that the product meets all the requirement of the customer, the quality of work is bound to be good. A person working with dedication voluntarily verifies the quality and ensures it to meet customer requirements, instead of just adhering to the process sheet given to him. He would not wait for the quality inspector to come and verify the quality. It may be checking the hank of sliver in draw frame or count of yarn in spinning or the pattern and ends and picks on loom.

Top management should take the primary initiative to keep the employees motivated in achieving the work quality. The systems developed should ensure that the stress is reduced at work. A good system designed with judicial thinking and implemented with proper empowerment, and whole hearted involvement of people can lead to achieve good work quality.

In today's high-tech and fast-paced world, the work environment is very different than it was a generation ago. It is common for a person in the textile or garment industry to change jobs for six to ten times in his or her lifetime, in technical as well as non-technical area. It is now rare for an employee to stay with a single company throughout his or her entire working life; whereas in companies such as Bombay Dyeing, Gokak, Madura, Century, Gajanana Mills Sangli, etc., which have completed over 100 years of successful working, we could see people working for generations in the same mill. As in the present economy there are more opportunities for people to work in different fields, employees are often willing to leave a company for better opportunities. Therefore, companies need to find ways not only to hire right qualified and competent people but also to retain them.

Now-a-days, most of the employees feel they are working harder, quicker, and for long hours than ever before and often complain of stress at work. Work-related stress can lead to lack of commitment to the company, poor productivity, and even leaving the company; all of which are of serious concern to management. Many employees work for 3–4 hours extra after their scheduled time, or come on weekly holidays and also bring work at home on a regular basis. With the Internet, laptops and mobile phones, it is now easy for people to work from any place, so get equipped with work even at home or even when they are on holidays.

1.2   Quality in work and work life

To get the best quality product, we need to have good quality raw materials, machines that are involved, people handling the operations, technology adapted and systems adapted. Good-quality raw materials can be procured by paying higher price, good machines can be procured by spending more, good people can be attracted by higher salary, good technology can be identified by taking help of experts in the field, but good systems cannot be borrowed. Good systems are to be developed in the organization by team work of management and employees, and should be sustained by respecting the system by top management. If top management respect the systems developed, the other employees are bound to respect and follow it.

Let us have a quick round of people working in the textile and garment industry. The mills and garment factories are facing short of skilled labour, but are not ready to treat the available skilled workers as a part of their industry. In number of mills, especially in south Gujarat and some other parts of North India, the workers are not appointed as regular workers, but are engaged as contract workers or cash workers on daily wages, as the mills do not want to have risk of making them permanent. These workers are removed from the job and again reappointed so that they cannot claim gratuity or any leave benefits. The workers are normally from U.P., Bihar, Orissa and Rajasthan. They leave their families at their native and come for work. The wages given to them is so low that they cannot afford to bring their family and settle here. They like to visit their home town once in a year for one month and hence are ready to work without taking any weekly holiday, and also work 12 hours a day. They are not paid any extra wages as specified by the Indian Factory Act, and weekly holidays are also not given. When these workers go to their native place, they are on loss of wages. The managements do not treat workers as an integral part of the company; hence, the workers also do not have any attachment to the company they are working in. They are more loyal to the person who brought them and not to the company. If that person leaves the company, all the workers and staff brought by him also leave together and the company suffers. Although salary and perks are lucrative to senior staff, they are also not willing to bring their family and settle as they feel insecurity in their job. They are always in the lookout for another job with the more security.

The mills in their attempt to reduce the production costs are neither bothered in the safety systems to be followed, nor in the welfare measures to be taken. The owner keeps all the powers with him; and even the general manager or the CEO cannot decide the raw material required or the system to be implemented. They never get the quality demanded by their customers and the staff and workers are blamed for this. The marketing team brings orders without understanding the company's capabilities in terms of quality and delivery and insists the workers to deliver. The people are always under stress to deliver on time and hence are forced to work extra hours. As a result, a textile technician working in such a company would not recommend anyone to study textile technology and join a textile mill. Neither the employees are happy nor the management. As a result, number of mills close down and new mills start with different name at different location. The technology in textile manufacturing is so much advanced that the quality can be achieved without much effort, and skill of workers needed is not as much as it was in earlier days; but even then industry is not achieving the quality aimed.

Quality in work can be achieved only when quality in work life is achieved. We have number of activities at home relating to our family, which we normally refer as our personal life. We spend more time in the factory we are working as compared to the time spent with family. We work both at home and at factory, but we are free and relaxed at home and always with tension and stress at mills. We have a feeling that our home is ours, but the same feeling is not there with the mills we work; although we are dependent on the mills to earn our bread. Why one does not feel as he is a part of the company and enjoy the work in the same way he enjoys with his family members? What is stopping us?

If we accept the industry as a part of our life, just like our home and take initiatives to develop it, the industry will prosper and we can be happier than what we are today. Our life, wherever it is, should be qualitative. There should be quality in work life, so that quality in work is guaranteed all the time.

A QMS expert, Imran Ahmed Rana, in his article titled "Quality Management System, Human Behaviour and Business Excellence," observes that organizations' social system is comprised on complex sets of human relationships interacting in many ways with each other and to the outside world. Besides belonging to the social and cultural settings into which people are born, in organizations people voluntarily join groups based on shared work practices, habits, beliefs, interests or knowledge levels. Memberships in these groups influence employee work traits, behaviour toward seniors, quality of work habits and perception about themselves and others. Consciously or subconsciously, these groups impose expectations and rules on its members. Rana suggests for arrangement for informal/formal coaching of these groups on organizational goals in the light of changing customer needs, so to extract positive and aligned work behaviour.

1.3   Objectives

Everybody does a work for some purpose. We are brought to this earth by God for some purpose; and we like to live on this earth for some purpose. If the purpose is achieved we shall be happy and call ourselves successful or winners. The purpose for which one works is called as his "mission". Each individual has some mission for which he works. Similarly, every organization has a mission. If we need to achieve quality in our work, we should be clear about the mission of company.

How can we say that our purpose to work is served? We need to have some standards to check our performance and achievements. Once we know what we have accomplished, we can plan our activities to achieve the remaining objectives. Objectives can also be termed as the measurable targets. We should identify the objectives of our work quality.

The objectives of work quality management is to achieve the required company objectives, while keeping the people involved including the stake holders happy at all levels. The achievement of work quality can be measured by various ways as listed below.

(a) Delighted customers because of timely supply of quality goods and services

(b) Delighted customers because of prompt and quick response to their enquiries

(c) Happy employees with no attrition, no absenteeism and no grievances

(d) Employees are willing to take higher responsibilities and referring their friends and family members to join the company

(e) No expenses incurred on legal cases in the court, no notices sent by any of the legal or statutory bodies for not meeting the compliance to any of the legal requirements

(f) No loss of production due to reasons such as breakdowns, lack of programs, poor working, short of materials, lack of clarity in instructions and so on

(g) Reduced cost of manufacturing

(h) Increased sales turnover per employee

(i) Increased sales turnover per rupee spent in marketing

(j) Reduced rejections

(k) No delay in delivery of any order

(l) No material returned back from market for any reason

(m) No complaint or grumbling from market not only regarding the product quality but also due to other reasons such as poor response, delay in transportation, discrepancies in sales documents, shortage or excess supply, higher price, etc.

(n) Zero accidents
(o) Zero fires
(p) Zero overtime
(q) Highly proactive in getting information from the market and changing the strategies and products to meet the changing requirement, and never blame recession for the losses, if any.
(r) Always being on toe-to-face competition and continuously stretching targets
(s) Being a benchmark for others and so on

1.3.1   Understanding the objectives and importance of work

Before doing any work, one should understand clearly its purpose and what is to be achieved. The purpose or the 'mission' is more important and the objectives to be accomplished come next to it. Starting without knowing the objectives may lead to wrong direction. In majority of cases, the subordinates do the work because it was told by their boss, but they fail as they do not know the purpose behind. If a worker in cotton blow room does not know the purpose of breaking cottons into small tufts before feeding to bale breaker, he shall either put more cotton without breaking the tufts resulting in jamming, or puts very less cotton making the machines to starve, and finally this would result into undesired lap density. A worker who does not know the purpose of a stop motion may inactivate it resulting in damages to machine in case of a breakage or meet with an accident. One who does not know the objectives of a safety bar in warping may meet with an accident. A sider who does not know the purpose of a door in the waste collection box may keep it open, resulting in poor suction and roller lapping in case of a breakage. The speedframe sider without understanding the purpose of separators put the separator down resulting in lashing of roving ends. In a production shed, the workers who do not know the purpose of fire extinguishers keep the material blocking the extinguishers causing inconvenience in case of a fire. The mill management who considers workers' training a useless job and sends untrained persons to machines may lead to poor efficiency of workers, poor quality, breakdowns and sometimes accidents. A supervisor, who does not know the purpose of a record, may enter the details improperly, and could not analyse it later. The tester, who does not know the purpose of a product, goes on checking lots of unwanted parameters, while leaving the critical ones required. A supervisor, without knowing the purpose of supervising, will simply take rounds without observing the critical points that need attention. The companies, without understanding the purpose for implementing quality management systems, may document lot of unwanted things, while leaving the important things specific for their company. The chairman of a company without understanding the purpose of writing quality policy by own and leaves the job to a junior officer or a consultant, will never be able to run the company as per his expectations. The company that does not know the importance of documentation by the men on job appoints a consultant to write procedures and manuals.

In number of companies, it is seen that some tasks are treated as very important and some are ignored. In a textile mill, each and every task is important. For instance, sweeping, if not done properly, shall result in poor products with more defects. If lavatories are not maintained well, workers may fall sick and their efficiency comes down. If workers are not allowed to take food and rest on time, their efficiency comes down and products will be defective. When the purpose of packing is not clear, one may not do the required packing and led the company and customer suffer. When the purpose of extracting comber noils is not clear, one may extract more noils and make the yarn costly.

Installing up the machines that do not fulfil the requirement always leads to undesired results. For example, if a textile mill installs the latest high speed air jet looms that work at 800 PPM speed, and runs fancy shirting with 500–1000 m length on beam, the machines would stop majority of the time for beam gaiting as the length of beam is very less. If running fancy shirting is the objective, then slow speed looms would have been more advantageous. Some process houses install high speed rotary printing machines but may not be able to get orders of long length on a single design with same colour combination; hence machine will be found stopped for changing the rollers. In the same way high speed warping machines are installed for running short length orders, these will get stopped for creel changing, and the mill may not get any benefit of high speed machines.

Need for departmental planning and jotting down the objectives is very important. It is seen that even if employees possess satisfactory level of technical competence, but still a wide performance gap may exist which may prove to be a hindrance in achieving organizational goals. Planning and establishing objectives reflects the behaviour and culture of the organization. There is a need for incorporation of this behavioural approach to management systems. Organizations can induce behavioural applications to their quarterly/annual planning and goal setting processes. At the onset of this process, important positive and negative work behaviours should also be brainstormed and recorded. To execute their goals, companies should align management system with behavioural approaches to work which shall help in improving work quality.

1.3.2   Dividing the work into small activities and defining their objectives

We should remember that an ocean is made up of small drops of water. All materials are made by small atoms and elements. The objectives are also made up by a combination of small steps and targets. Hence the objectives, if are to be achieved, need to be broken into small targets, and allot it to small sections or individuals for achieving. When we talk of an organization, it is needed to break the company objectives into departmental objectives, and then to sectional objectives. Finally, the targets are to be fixed for individuals, and they should be empowered to work for achieving those targets without any intervention. If all the individuals achieve their targets, the company target shall be achieved. Action plans to achieve the targets are to be prepared with time bound schedules. A sincere effort in implementing the action plans is required if the targets are to be achieved.

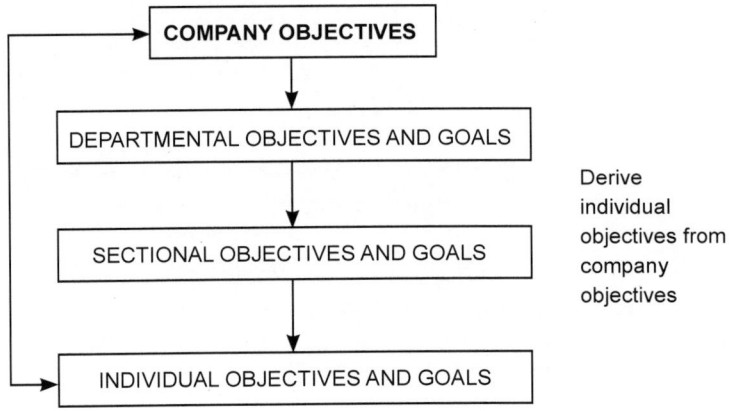

Let us take an example of loom shed where management wants 80% efficiency from the present 70% efficiency. The first activity shall be to find out the reasons for getting 70% efficiency and breakup of the loom stoppages for 30%. If yarn break is the reason, then there is a need to improve the raw materials after ensuring the size pick-up and humidity maintained are normal. If the time lost for beam gaiting is more, then there is a need to analyse the activities of beam gaiting and to find out where exactly the time is lost. If the weaver is untrained and taking more time to attend the breaks, there is a need to arrange his training. Analysing the errors and then taking the decision takes more time, so every activity needs to be attended uniformly. The yarn sourcing person should be advised to source better yarns and at the same time the weaver should be trained to work faster by understanding priorities when more than one loom is stopped at a time. Care should be taken to check that all materials are brought and kept ready before a loom stops for gaiting or maintenance. The process of beam gaiting involves numerous activities, so the distribution of tasks should be done properly to avoid interference. Along with beam gaiting, preventive maintenance of the loom also should be done. It may be cleaning, oiling, replacing of some parts and so on. If the objectives are defined to the team properly, then the work can be done effectively and efficiently.

The efficiency can be increased by proper coordination among the workers handling different tasks, and thus each one contributing equally in achieving the targets with interest and commitment. The work quality of workers, supervisor, manager and the management are all important in achieving the loom shed efficiency.

1.3.3   Developing measures and methods for monitoring and control

Breaking the task into various activities helps to monitor easily the performance of each activity. Therefore, setting up the standards and matching the performance against these standards help to locate and rectify the errors. For example, beam gaiting involves the below activities:

1. Removing run-out beam
2. Removing drop wires
3. Removing healds
4. Removing reed
5. Cleaning the loom and parts to be oiled
6. Oiling and greasing the loom
7. Setting the width/stroke as per the reed space required
8. Mounting new beam with drop wires, healds and reed
9. Setting loom for new beam design parameters without defects
10. Starting up the loom
11. Taking sample cutting and going for production

What should be the standard time for removing the beam? Whether removing beam from the loom and putting on beam trolley can be done by one man? It depends on the type and width of loom, and on the dimensions and weight of the beam. One should have studied the time taken by different teams and arrive at an achievable standard time. The people working should have clarity of the method of doing work.

How to remove the drop wires? Where to keep them? How to ensure different sizes and types of drop pins are not mixed up? How much time is needed to remove the drop wires? How many people have to work? Go on asking questions to yourself and find answers.

How do we say that the beam gaiting activity was done well? We need to have some parameters that can be measured and verified. Normally, people consider the time taken from stopping of the loom to the starting of the loom after gaiting and setting the quality. If gaiting is not done proper, the warp breaks shall be more on the loom. While bringing the drawn beam to the loom, if handling is not proper, we may get more ends broken and time taken to re draw the ends may take more time. The quality of work at the first instance is very important.

Go on asking questions at each stage and come to a logical conclusion and decide the methods of monitoring and control. Find out the factors responsible for success or failure of an activity. Once all the factors responsible for successful working are analysed, one can prepare a checklist and verify those parameters and ensure that the work is perfect, or meet the expectation.

Once the method of doing a work and the controls to be exercised are understood clearly, it can be documented as a procedure. The documented procedure helps all in the organization to work in the same way and get uniform and consistent results. It's not necessary that one should document procedures only when going for ISO 9001. Documenting and implementing the activities accordingly improves the quality of work.

1.4   Procedures

Establishing the procedures and working according to these help to accomplish the objectives effectively and efficiently. If procedures are not defined, there are chances of failures. Majority of the problems in the textile industry are due to not defining and following the procedures accordingly. Most of the people take shortcuts to get quick results and hence end up in a mess. The main culprit is the top management, who sometimes even after knowing the procedure to be followed, insist the subordinates to bypass the procedure and do the work. Bypassing the authorities, avoiding quality inspection, not analysing the vendor capabilities before purchasing materials, not verifying the minimum competency level while recruiting people for job, not recording the extra hours worked and not paying the workers as per the law, promoting a person without verifying the competencies and neglecting other people in the organization eligible for promotion, not installing safety systems and not checking safety systems etc., are some examples. All these lead to poor work quality.

To make the organisation function in an effective manner, defining the procedures and documenting these with clarity and religiously is very essential. People involved should be educated and convinced to follow the procedures.

In number of textile and garment units we can see the procedures and policies documented nicely by employing a consultant, but are not being followed. The people on shop floor are not at all aware that such procedures are written, and the top management is not insisting people to work as per procedures; even they are not working as per the procedures documented. The manuals and procedures are kept clean only for showing to auditors and customers. The quality of work can never be achieved unless the procedures are implemented whole heartedly. How to make management to implement the policies and then convince the staff to implement these is the major challenge in front of the industry.

1.4.1   Difference between procedures and instructions

Procedures – These indicate the way of doing a work to achieve a specified objective, the people involved in it, the responsibility of each in making the activity fruitful, the factors to be controlled and checks to be done, and recording and reporting of the activity. Developing procedures is the first step of getting success. If procedures are appropriate to the activity and the objective, we are certain to get the results provided if we follow these religiously. The procedures need to be established before documenting and ensured to be followed religiously. The interactions between activities in a process should be clearly explained in the procedure.

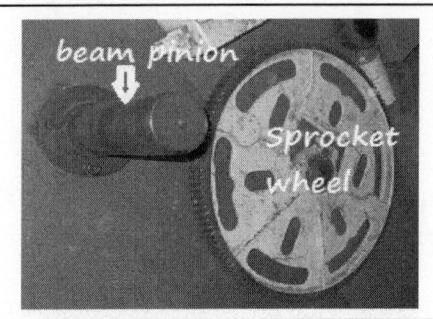

Dismantle the empty beam, remove beam pinion and beam sprocket wheel.

Work instruction – Numbers of small activities are involved within a procedure. The person doing that activity should be clear of that work. For example in the activity of beam gaiting, the activity of removing the beam includes removing beam pinion and beam sprocket wheel, taking the beam out and keeping on the empty beam trolley and transporting it to a specified location. The worker responsible for doing this work should be clear about the activity. He should know what the beam pinion is, how to remove it, how to remove the beam sprocket wheel, tools to be used to do these jobs, the method of taking the beam out without hurting himself or damaging the machinery, the method of keeping it on the beam trolley and then moving it to specified location. This should be documented with pictures and given to the worker so that he can refer and do the work. This is called as work instruction.

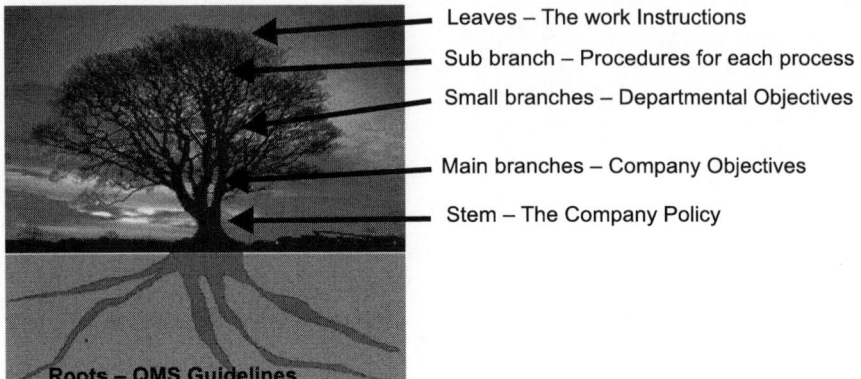

Leaves – The work Instructions

Sub branch – Procedures for each process

Small branches – Departmental Objectives

Main branches – Company Objectives

Stem – The Company Policy

Roots – QMS Guidelines

The QMS guidelines should be made the foundation for documenting the policy, objective, procedures and instructions. Considering tree as an example, QMS could be called as roots and policies could be the stem. There can be one strong policy and not hundreds. The company objectives are derived from the policy as the main branches bifurcate from the stem. The departmental objectives are derived from the company objective, and procedures are written to achieve departmental objectives, as branches are subdivided in a tree. For each procedure, there may be number of work instructions like leaves of a tree.

1.4.2   Developing procedures to achieve objectives

To do any work successfully, we should be clear about the method of doing work, the sequence of activities, the breakup of activities and assigning these to different members of the team, the controls and checks at appropriate places, coordination of activities of individuals and combining them and finally reviewing and reporting. Whatever may be the experience and knowledge a man has, it is not possible all the time to remember everything and follow all the steps. In order to avoid the misses, and also to help subordinates to develop, it is necessary to document the procedures and insist people to read it on regular basis religiously. In number of mills, it is seen that documents are kept in lock and key and not given to people actually doing the work. The procedures are written by a consultant as the people working on the spot (and sometimes the management also) claim/feel that they do not have time to write the procedures or do not have competency to write. Even the HOD does not try to read and explain to his subordinates. A junior assistant is given the responsibility of maintaining all the documents neatly in a secured place so that they can be presented to auditors during audits. With such system, improvement cannot take place, and the documented procedures have no meaning. This cannot improve the work quality.

Any document of a procedure should first explain the purpose and objective of doing that activity. It should clearly specify the result expected out of that activity. After specifying the purpose and objectives, we need to define the scope of operations, the area in which these operations shall take place, the people responsible for doing different activities within these operations.

Once the purpose, scope and responsibility are spelled out, it is better to explain the importance and the principles of operations, and the technical information in case of operating machinery so that people can be clear about various functions available in the system. Afterwards the work to be done should be explained step by step so that the people involved do not miss a single step.

In the procedure write clearly as what is to be done, when it is to be done, where it should be done, how it is to be done, and who should do it.

It is advisable to highlight the important steps at the end in the form of Do's and Don'ts, so that people shall be alert. In some companies, there is a system of writing frequently asked questions (FAQs) after the procedures, so that the implementers can think appropriately and actively participate in modifying the procedures as required and use their creativity in doing the work.

All procedures documented need to be scrutinized by the head of the section that is responsible for giving the results and to be approved by the chief of the operations. The approved versions of the procedures are to be circulated among the people doing the work for reference. All people involved in the activity should have an access to the procedures written for their activities. The people should be empowered to do the work as per the procedure, and no one should interfere or insist taking a short cut or bypass a step to get the work done fast. This only can improve the quality of work.

1.4.3   Preparing instructions and checklists to implement procedures

Once the procedure is written down, identify the elements of the procedure. For example, in the beam gaiting exercise for a rapier loom, the actions are

1. Dismantle the empty beam
2. Remove beam pinion and beam sprocket wheel
3. Transport empty beam to the specified location
4. Dismantle stop motion device and drop-pin bars
5. Dismantle heald shafts and brackets on both the sides
6. Dismantle the reed of the run out beam loom and transport it to the reed stock location with loom no. and design no. of run out beam
7. Clean the loom thoroughly using compressed air pipe
8. Clean the bushes
9. Oil/grease at all the specific points depending on the type of loom as recommended
10. Attend required maintenance work in the loom
11. Transport the required beam to the back of the loom to be gaited in the carrier
12. Disengage the new beam slowly and keep it at the beam fitting position
13. Keep the heald shafts properly on the bracket
14. Place the drop pin bars on the specified positions
15. Tie the heald shaft together with a cord on both sides
16. Replace the plain lease bars in the new warp beam with the serrated lease bars of the loom
17. Set heald shafts in the bracket
18. Check that front shafts are kept for selvedge ends
19. Assemble the reed, place the warp on the back rest
20. Open bunches of ends tied into knots in front of the reed and pull them
21. Knot the bunch of end of new beam with the bunch of old end on loom
22. Dress and set warp so that ends are parallel and placed evenly across the width of warp
23. Adjust the drop pins across the ends of beam
24. Insert the reed at specified position correctly
25. Pull the warp ends through them
26. Set the beam tension as required by slackening the beam and setting in computer
27. Set the weft insertion mechanism as per new design of the beam
28. Set the rapier for new reed space if previous reed space setting is different
29. Dismantle the grippers both left and right
30. Dismantle the opener

31. Insert the reed
32. Pull the ends through them
33. Set stoke of gripper guide as per reed space required for new beam
34. Set the gripper keeping the right degree
35. Check the tension in warp and adjust it
36. Set Leno winding on both sides
37. Check and place the required weft package in the creel
38. Draw the weft using hook through the guides and fingers
39. Set the finger selection as per weft feeding pattern required in the computer
40. Finally start and run the loom operating inch button

Now decide on which of the activities you need to write work instruction so that the worker cannot commit a mistake. The work instruction to be written depends on the level of maturity and training given to workmen. If the workmen are well trained and qualified, we can reduce the number of instructions and just write Do's and Don'ts. Following is an example of Do's and Don'ts for beam gaiting in a rapier loom.

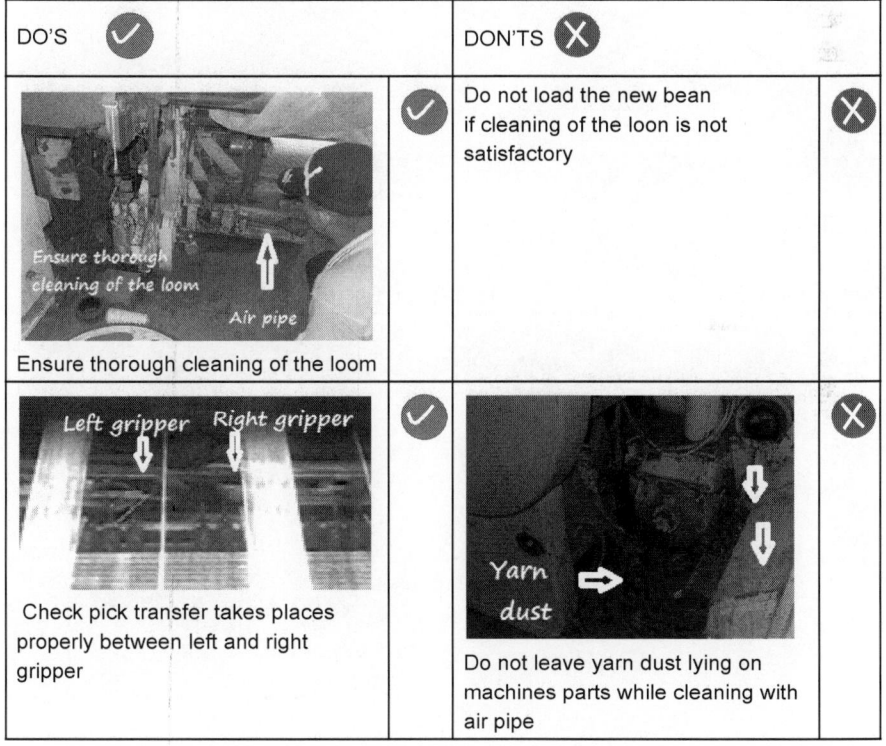

DO'S ✓		DON'TS ✗	
	✓	Do not load the new bean if cleaning of the loon is not satisfactory	✗
Ensure thorough cleaning of the loom			
Left gripper Right gripper	✓	Yarn dust	✗
Check pick transfer takes places properly between left and right gripper		Do not leave yarn dust lying on machines parts while cleaning with air pipe	

Identify the missing ends in pattern with the ends pasted with gum tape on warp sheet

Do not run the loom if missing ends is observed

Set the cutter by adjusting the nuts as required for proper cutting of waste selvedge

Do not run the loom without correcting warp floats due to improper programming

Check and set heald shafts properly

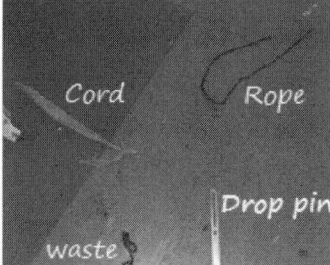

Do not throw drop, pins, cords or waste on floor

Keep the cut run out warp in a plastic crate	✓	Do not put the cut warp of run out beam on the floor ⬇	✗
Temple ⬇ Ensure correct setting of temple	✓	Do not run the loom if the temples are not properly set	✗
Serrated drop pin bars Check that plain drop pin bars are replaced with serrated drop pin bars before starting the loom	✓	drop pins stop motion device ⬇ Do not forget to fit the warp stop motion device	✗
  Check and set cutter as required	✓	Do not run the loom if cutter is not set for correct waste selvedge cutting	✗
Warp beam should be accurately fitted in to the brackets	✓	Do not run the loom if the reed is old, damaged or rusted	✗

Use mask to protect yourself from inhaling dust	Do not dust the machine without mask

1.5   Time management

We need time to do any work. Time is a resource, which is very regular. There is no deviation. It comes in time and goes off when time is over. If we do not use it in time, it is our mistake. Unlike other resources we cannot store the time to use it when required. If we miss it, we cannot get back. Hence we need to learn how to manage our works during the time available. People refer it as "time management", but actually no one can manage time. It is "managing the activities within the time." If all the works are managed in the allotted time, you are bound to achieve work quality.

We all know that the time allotted for a minute is 60 seconds, for an hour is 60 minutes, for a day is 24 hours, for a week is 7 days, and for a year 365¼ days, which are all fixed and cannot be changed. This resource comes to us without fail.

1.5.1   Listing the activities

List out the activities to be done and prioritise these on the basis of importance. Allot the time for each activity by making sequential analysis. Eliminate the activities which are not essential and are not adding value. Do not deviate from the planned programme. We should take one task at a time and complete it and get the satisfaction of completing it and achieving the goals. After completing one task, take the next task. In a number of times it may not be possible to follow the timing exactly as per the plan, but an attempt to plan and follow helps to plan the activities more effectively. Following is an example of a weaving shift in-charge how he had made his daily working schedule.

Timings	Shift in-charge (weaving)
7.30 to 7.45 AM	Taking round with previous shift in-charge
7.45 to 8.00 AM	Understanding the situations and making rough plan. Discussing with supervisor
8.00 to 8.30 AM	Verifying the production and quality of previous shift and planning
8.30 to 9.00 AM	Taking round with VP in Somet rapier looms
9.00 to 9.30 AM	Taking round with VP in Dornier air jet looms
9.30 to 10.00 AM	Following up for the materials
10.00 to 10.30 AM	Following up for housekeeping
10.30 to 11.00 AM	Taking round to assess the production till time
11.00 to 11.30 AM	Taking periodic quality round
11.30 to 12.00 noon	Following with stores and warehouse for the materials
12.00 to 12.30 PM	Taking round of the loom shed and noting down picks produced
12.30 to 13.00 PM	Taking round of the loom shed – observe house keeping
13.00 to 13.30 PM	Taking round of the loom shed – observe the wastes produced
13.30 to 14.00 PM	Lunch
14.00 to 14.30 PM	Taking round of the loom shed and verify picks produced on each loom
14.30 to 15.00 PM	Taking round of the loom shed
15.00 to 15.30 PM	Taking round of the loom shed
15.30 to 16.00 PM	Taking round with next shift in-charge
16.00 to 16.30 PM	Checking production reports and reporting to HOD
16.30 to 17.00 PM	Meeting VP and explaining the highlights of the day

While making such planning, one needs to discuss with the interacting staff such as his superiors and juniors and adjust the time in such a way that it is suitable to all. Following is an example of time table made by VP (weaving) who had under him weaving preparatory, and two sections of looms, one rapier and other air jet looked after by different heads.

Timings	Vice President (weaving)	Head of rapier looms	Head of air jet looms	Head of preparatory
7.45 to 8.00 AM		Understanding the problems and conditions by discussing with outgoing staff	Understanding the problems and conditions by discussing with outgoing staff	Understanding the problems and conditions by discussing with outgoing staff

Cont...

Timings	Vice President (weaving)	Head of rapier looms	Head of air jet looms	Head of preparatory
8.00 to 8.30 AM	Check the mails received and the schedules of the day	Taking round in the weaving section for materials and housekeeping	Taking round in the weaving section for materials and housekeeping	Taking round in the preparatory section for materials and housekeeping
8.30 to 9.00 AM	Taking round of rapier section	Taking round along with vice president	Checking productions, quality and housekeeping by taking round	Taking detailed round in winding area for house keeping
9.00 to 9.30 AM	Taking round of air jet section	Attending to points shown by VP in his round	Taking round along with VP	Taking detailed round in winding for process parameters
9.30 to 10.00 AM	Discussions relating to efficiency, production and points observed during rounds	Taking part in discussions with VP	Taking part in discussions with VP	Taking detailed round in warping and checking the stoppages and efficiency
10.00 to 10.30 AM				Taking detailed round in sizing and verifying complaints from weaving
10.30 to 11.00 AM	Grey folding quality meeting and housekeeping rounds	Taking part in quality meeting along with vice president	Taking part in quality meeting along with vice president	Taking part in quality meeting along with vice president
11.00 to 11.30 AM				
11.30 to 12.00 AM	Taking rounds in yarn godown, Winding, warping and sizing.	Attending to points discussed with VP	Attending to points discussed with VP	Taking round along with vice president
12.00 to 12.30 PM		Attending to points discussed with VP	Attending to points discussed with VP	
12.30 to 13.00 PM	Discussing on the issues relating to preparatory	Attending to points discussed with VP	Attending to points discussed with VP	Taking part in discussions with VP

Cont...

Timings	Vice President (weaving)	Head of rapier looms	Head of air jet looms	Head of preparatory
13.00 to 13.30 PM	Visiting finish folding, understanding the problem relating to weaving	Visiting finish folding, understanding the problem relating to weaving	Visiting finish folding, understanding the problem relating to weaving	Instructing the in-charges and supervisors relating to the problems found
13.30 to 14.00 PM				Visiting finish folding to understand the problems related to preparatory
14.00 to 14.30 PM	Visiting grey warehouse	Instructing the in-charges and supervisors relating to the problems found	Instructing the in-charges and supervisors relating to the problems found	Instructing the in-charges and supervisors relating to the problems found
14.30 to 15.00 PM	**Lunch**	**Lunch**	**Lunch**	**Lunch**
15.00 to 15.30 PM	Check the mails and the schedules completed of the day	Check the mails received and the schedules completed of the day	Check the mails received and the schedules completed of the day	Check the mails received and the schedules completed of the day
15.30 to 16.00 PM	Discuss the issues with others as needed	Take round of the section	Take round of the section	Take round of the section
16.00 to 16.30 PM	Discuss the issues with others as needed	Interacting with weaving shift in-charges regarding the production and efficiency of the shift	Interacting with weaving shift in-charges regarding the production and efficiency of the shift	Interacting with preparatory shift in-charges regarding the production and efficiency of the shift
16.30 to 17.00 PM	Reviewing the performance of the shift of loom shed	Reporting the shift production and efficiency to vice president	Reporting the shift production and efficiency to vice president	Following up for production
17.00 to 17.30 PM	Attending PPC meeting	Attending PPC meeting	Attending PPC meeting	Attending PPC meeting
17.30 to 18.00 PM	Planning for next day	Planning for next day	Planning for next day	Planning for next day

Cont...

Timings	Vice President (weaving)	Head of rapier looms	Head of air jet looms	Head of preparatory
18.00 to 18.30 PM	Meeting CEO and discussing the issues	Meeting CEO and discussing the issues	Meeting CEO and discussing the issues	Meeting CEO and discussing the issues

Above type of rough planning helps others in the section also to plan their activities in line with the activities of their bosses, and can work with cool and calm mind.

1.5.2   Identifying the essential, urgent, important, not-so important and avoidable

First group the activities as 'urgent' and 'not urgent'. The urgent works are to be given preference. Similarly group the works as 'important' and 'not important'. Again important works are to be given preference. By combining both, we get 4 groups, which can be explained as urgency – importance matrix. The groups we get are as follows:

1. Urgent and important
2. Important but not urgent
3. Urgent although not important, and
4. Neither important nor urgent

The Group 1 works get first priority, and no one objects them. The works in Group 3 are urgent, and hence, somehow or other they also get the priority. The Group 2 activities, even though they are important, shall be getting postponed as they are not urgent. By this attitude, they shall become urgent after some days, and we will not be able to accomplish it as we have not planned. It is the normal tendency among many people, that they do not start the work unless it becomes urgent. In fact, it is easy and convenient to plan the activities when they are not urgent. Hence by giving priority to this group of work, the urgent works automatically reduce.

	Group 3 Urgent but not important	**Group 1** Urgent and important
Urgency	**Group 4** Neither important nor urgent	**Group 2** Important but nor urgent

Important

If we are always busy in doing urgent works, it means there is something wrong in our management. Good managers keep time reserved for the Group 2 works. This includes planning, team building activities, training, SWOT analysis, review of activities, etc.

The Group 4 activities are also called as "time wasters", but we cannot leave this. In fact people love to do this activity. A number of reasons are given for doing such activities, as they come under the category of stress relievers, mind relaxers, entertainers, etc. They might not be important or urgent considering our goals or business, but are required to be done due to one or the other reasons. This includes chitchatting with friends, commenting on the activities of others with whom we have no relation or connection, just to pass the time, relaxing, going on tours, viewing movies, watching sports, reading novels or literatures, taking part in entertainment activities, etc. There are other works in this group such as working for obligation, doing something to show that we are great, doing something for the sake of records, spending or doing something to maintain our prestige, etc.

1.5.3   Allocating time

Prepare the log of activities with time, and rank them in the category or group as shown in the urgency – importance matrix. Let us see an example of time log written for an office employee.

Time log of an office employee

Time	Activity	Category
5.00–5.30 Hrs	Getting up, brushing teeth and attending natural calls	1
5.30–6.30 Hrs	Morning walk	2
6.30–7.00 Hrs	Reading newspaper	2
7.00–7.30 Hrs	Taking bath and dressing up	1
7.30–8.00 Hrs	Taking breakfast	1
8.00–9.00 Hrs	Driving – going for work	1
9.00–9.30 Hrs	Checking mails and replying them	1
9.30–10.00 Hrs	Answering calls from friends	3
10.00–11.00 Hrs	Organizing the files and registers	2
11.00–11.30 Hrs	Taking tea and chitchatting with colleagues	4
11.30–12.00 Hrs	Preparing report to be shown to boss	1
12.00–12.30 Hrs	Waiting in front of office of boss	4
12.30–13.00 Hrs	Submitting the report and discussing	1

Cont...

Time	Activity	Category
13.00–13.30 Hrs	Correcting the statements as advised by boss	1
13.30–14.00 Hrs	Lunch break	1
14.00–14.30 Hrs	Waiting for boss to give corrected papers	4
14.30–15.00 Hrs	Submitting reports and discussing	1
15.00–15.30 Hrs	Correcting the statements again	1
15.30–16.00 Hrs	Taking signature of boss	1
16.00–16.30 Hrs	Photocopying the papers and giving originals for despatch	1
16.30–17.00 Hrs	Taking tea and chitchatting with colleagues	4
17.00–17.30 Hrs	Attending to visitors, personal guests of boss	3
17.30–18.00 Hrs	Chitchatting with colleagues while closing the files	4
18.00–19.30 Hrs	Driving – going back home	2
19.30–20.00 Hrs	Freshening up and taking Tea	3
20.00–20.30 Hrs	Watching TV serial	3
20.30–21.00 Hrs	Watching news at TV	2
21.00–21.30 Hrs	Reading books	2
21.30–22.00 Hrs	Dinner	1
22.00–22.30 Hrs	Listening music	2

The activities of a particular day are planned with a time gap of 30 minutes. The categorisation is done as per the perception of the man who does the work. Others might be put in different categories. For example, an employee categorises waiting for his boss in Group 4, as he feels it as neither urgent nor important. The boss might say it as important, as he cannot wait for others, but others have to wait for him. He feels he is busy. By this attitude, he makes his subordinates inefficient. Answering calls of friends is grouped as no. 3 as he felt them as urgent but not important. If we ask the friend, who called on him, would say that it was urgent and important, that is why he called him. It means we allocate time to number of works on urgent basis, even though it is neither urgent nor important for us, but because it is urgent or important to our stakeholders. A good manager politely refuses such activities, whereas some refuse in a harsh way. The one refuses politely shall be able to win, whereas the one adopted a harsh method loses cooperation from others. The one who does not refuse cannot find time for doing own works, and hence cannot win. Telling "NO" when we do not want to say "YES" is an art, and one has to practice it.

1.5.4 Time log

Prepare time chart, i.e. expected time for doing an activity and at what time the work should be done for the routine activities, and mark the priorities. You should ensure that you work within the time frame and the priority tasks are done first. After completing the work, mark the actual time taken. For example, you had allotted 10 minutes for an activity but actual time taken was 9 minutes or 12 minutes. You mark the actual time. At the end of the day, you can see what was the deviation from the planned, and what you should take as standard while planning next time. Wherever the time taken was more than the normal, analyse the factor contributing for the delay. We might have taken less time in the planning than what it should be, or some act might have consumed more time due to some reasons. For example, you have allotted 15 minutes for discussing a matter with your MD, whereas you spent more than 40 minutes, as MD was asking number of other questions not directly related to the topic you wanted to discuss.

Make an analysis of the delays over a period of time and find out what is not avoidable and likely to interfere in the work. Keep some allowance for such unavoidable tasks. For example, you cannot avoid if your MD is asking more questions from different areas, much away from the subject you wanted to discuss. These include delays in receiving materials, interference in works, changes in priorities, and delays in getting clear instructions, more time taken in discussions, power failures, and non-completion of work by team members in the same time and so on. Do not keep unnecessary extra allowance, as it will make you inefficient.

The time allotted by you should be such that your team can successfully complete the activities on time, but still have energy and enthusiasm and satisfaction of completing the work in time. They should feel the challenge. Without challenge, you will not get satisfaction in work.

1.6 Work management

While we talk of work management, we should start with defining the work and objectives to be achieved, planning the work to achieve the objectives that include planning the materials, developing methods, planning the sequence of works, allocating time for each activity, monitoring the activity and reviewing the results at appropriate intervals. Work management composes of quality management, time management and production management while ensuring that the safety and social requirements are met.

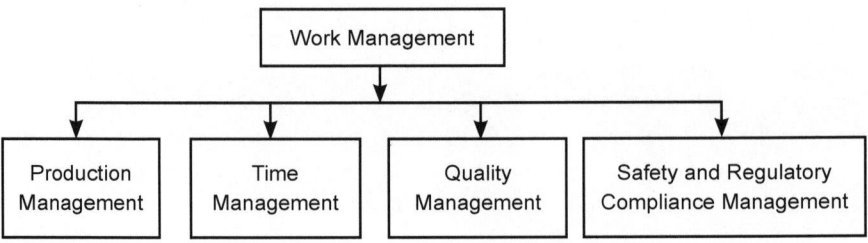

1.6.1   Planning for the materials

Planning the materials include not only planning for the quality of materials but also the quantity and the time at which the materials should reach the work spot; this saves from unnecessary stocking and the delay in production as well. Materials include not only the raw materials but also the accessories. It means we need to plan and monitor the complete supply chain. Any failure at any place in the supply chain leads to failure of the complete system or product.

Materials can be planned properly when we are clear about the wastes extracted, the rate of production, the lead time to procure the material, the quality level expected and the provisions we have for handling and storing the materials.

The quality level to be achieved becomes the main criterion followed, with the cost of materials for selection of the raw material. When the cost of raw materials is high and becomes impossible to price the product according to the affordability of the customer, one shall be forced to procure a cheaper available raw material, and the competency of the technical staff is decided on how best they will be able to realize the product with the given raw material. They need to develop innovative methods of salvaging good materials and remove unwanted from the raw materials, while setting the parameters to produce less rejections and wastes. The technicians are supposed to go on observing the process and eliminate unwanted and non-value adding activities to reduce the cost. The work quality of a technical person is decided by this.

1.6.2   Developing methods

The methods should be developed considering the existing raw materials, the infrastructure available, the level of people working on the shop floor and all types of shortcomings in the company. The method should be implementable by the concerned working for that. Too ideal system may not help as it cannot be implemented unless all are ideal. The method should be able to reduce tension and fatigue while working. Therefore while designing the methods ergonomics and movement between processes are also to be considered.

The culture of the people working plays a very important role. For example, the workers from north have habit of eating gutka and spitting wherever they are. They even do not bother whether it is a wall or a machine or the product being manufactured. In a number of cases it is seen that even the senior staff have the habit of chewing gutka and spitting where they are working.

Habit of spitting after eating gutka

The people working should respect the materials they are working with. In number of mills, especially in Surat area, it is seen that people just walk on the materials stacked, which are not covered. They do not bother to cover the materials when cleaning is taking place.

Poor handling of cones resulting in stains and mix ups.

The handling practices, housekeeping, stacking and transportations should be addressed in the working methods apart from handling of machinery. A poorly maintained machine cannot give result although the best raw material is supplied. In a number of north Indian mills, where the workers and staff are migrated from UP, Bihar, Orissa and Rajasthan, and are kept on contract basis in order to reduce the cost of manufacturing have the problems of poor quality, higher wastes and low productivity, as the workers are not working regularly and do not have a feeling of belongingness to the company.

	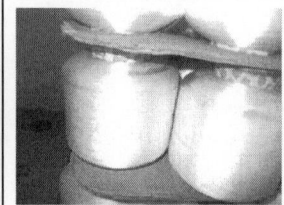	
A poorly maintained winding machine running on Polyester filament yarn	Poor handling of polyester filaments resulting in stains	Poor handling of bright filament yarns resulting in wastes

The mills must concentrate on retaining the workers and training them with good work methods if they are interested in real performance. The short-term policy of hiring people through contractors and keeping them on daily wage basis is one of the main reasons for poor maintenance, poor handling, higher wastes and damages. The workers also should develop the culture of respecting their job as it is providing them food and livelihood.

1.6.3   Planning the sequence of works

Planning the sequence of operations is the next task in developing methods for work management. We need to select the tasks that should be done first, and which could be done simultaneously. For example, a weaver handling eight looms needs to work simultaneously on all the eight looms. If there is a warp beak in one machine and weft break in another, he will choose first to attend the weft break as the machine can start faster, and then comes to attend warp break. In case of a warping machine, when the machine is working, the creel boy will start creeling the reserve creel for next lot, and he will come back in case of break or help required by the warper. If we take the example of beam gaiting, the beam drawing must have been completed much before the running beam runs out on loom. The drawn beam should be brought and kept near the loom, of course, with sufficient place to remove the old run-out beam and to take it away. All the team members must be ready at the same time as lifting the beam cannot be done by single man. The required no. of people should be available as per the requirement of work. Take the example of doffing in a ring frame: each doffer shall prepare his doffing basket with required number of empties separately, but when the machine comes to doff, all will be standing in their respective positions. Each one will start taking out the full bobbins and putting new empties in the spindles allotted to them and all will complete this work at a time. The machine will start and each doffer pieces the broken ends in the allotted spindles.

To plan the sequence of work, the planner should be clear about all the activities. This task cannot be assigned to a fresher, just because he is qualified from the university. One has to observe each and every activity with patience and understand the importance of that activity. In number of cases, an outsider will not understand the importance of the activity. For example, while dressing ends on a loom, the dresser will be straightening each bunch of warp ends number of times. A fitter will not make all the bolts tight in the first time itself; he will fit and tighten each bolt slightly and finally he makes all fully tight.

1.6.4 Allocating time for each activity

After understanding the sequence of works, write a PERT Chart (Project Evaluation and Review Chart) and the time required. Following is an example for a project. A is the starting point and M the ending point. There are four parallel activities referred as paths. The number of days for each activity is written by the side of the arrow.

Typical PERT diagram

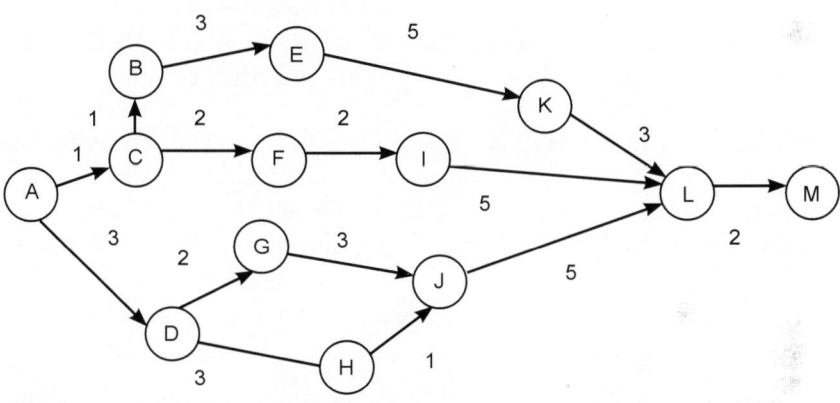

Start = A   End = M
Path 1 = A → C → B → E → K → L = 1 + 1 + 3 + 5 + 3 + 2 = 15 days
Path 2 = A → C → F → I → L → M = 1 + 2 + 2 + 5 + 2 = 15 days
Path 3 = A → D → G → J → L → M = 3 + 2 + 3 + 5 + 2 = 15 days
Path 4 = A → D → H → J → L → M = 3 + 3 + 1 + 5 + 2 = 14 days
Path 1 and 3 takes more time, i.e., 15 days, hence they are critical

By allocating the time for each task, problem can be detected and rectified easily.

1.6.5   Monitoring the activity

Doing the right work on right time is known as quality work. We need to know the critical points that need to be checked. For example, a doffer boy removes 100 cops in 1 minute whereas the other boy takes 1.5 minutes. By this figure, we think that first boy has handled the task effectively, but finally the output of work quality matters. The critical observation found that the first boy has broken 50 ends whereas the second one has broken 10 ends. The second boy finishes piecing in 1.5 minutes, whereas the first boy needs 7.5 minutes. Taking doff without breaking the ends is more important than just taking doff fast. Similarly at all places, doing a good work is more important than just doing fast.

Work monitoring, therefore, includes not only monitoring the time for each job, but also ensuring quality of work at each stage.

1.6.6   Reviewing the results

Did you get the results what you wanted? Check the output of work. How much you achieved and how much is lagging? Sometimes, you might have achieved better than your own expectations. In such a case, verify the factors that helped you to achieve better results. Work out whether you can make it as a permanent feature.

While reviewing the results, always make it a practice to compare not only with the target, but also with the trend: Are we moving up or going down or are irregular in our performance? Let us take an example of 3 ring frames running on 20s KWP. The following chart indicates the production in grams per spindle for 8 hours.

Production in 20s KWP in grams/spindle

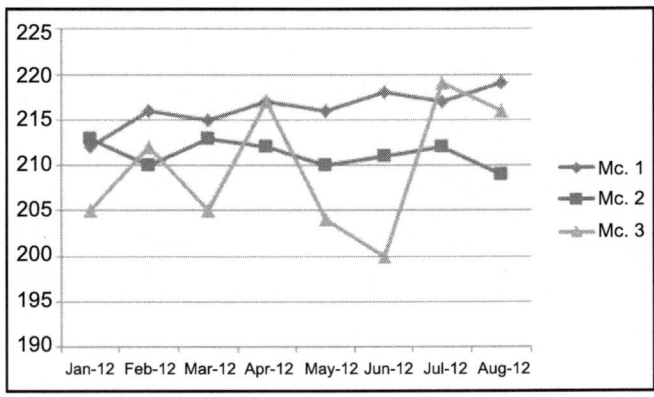

The machine 1 is showing gradual improvement, whereas machine 2 is at same level or slightly coming down. The machine 3 is erratic. Here we need to emphasize on machine 3 first and make it stable, and then check machine 2. Problem leading the machine 1 to gradual improvement also needs to be studied.

1.7   5-S concepts

There are a number of sub-processes in each process: some contribute directly to the success of the process, while others might hamper the process. It is essential to define the objectives of the processes very clearly, in order to have clarity while designing the process. One needs to understand what is required and not required. By eliminating the unwanted activities, we get time to concentrate on essential activities. Use of 5-S concepts in work process management, which was originally developed for managing housekeeping, can help in this. The concepts of 5-S are mentioned below:

Seiri → Sorting out

Seiton → Systematic arrangement

Seiso → Spic and span cleanliness inspection

Seiketsu → Standardizing

Shitsuke → Self-discipline and training

Sorting out the not-required materials and removing these from the work place is the first principle of 5-S. This helps in providing place to keep the required materials and to do the activities without hurdles. Unwanted materials, unwanted processes and unwanted persons create hurdles and problems. This reduces quality of work and spoils peace of mind. Similarly unwanted thoughts spoil our thinking process, and we might be pushed unnecessarily to unhappiness.

Unwanted materials occupy the place and make everything shabby, reduce efficiency and create irritation to mind.

For a task or the work we are doing, we need to categorize the wanted and unwanted subtasks involved in it. Eliminating the unwanted ones will reduce the cost and save time. This would also help to reduce the tension. After a close observation, we may find out that we do a lot of unnecessary tasks and incur additional costs. The concept of Seiri says that removing the scrap lets you to get space released for other materials. Move such materials to a common place, as these might be unnecessary to you but useful to other department. The common place for keeping these materials is called "Red Tag Area" or "5-S Museum". This can be located at departmental level or company level. In houses, it might be a store room or a dedicated cupboard. Store away the materials used less frequently and retain the materials frequently used near the work place. This will avoid confusions and mix ups, and improves the efficiency. Preparation of a time log, and recording the activities and grading these as per the importance can help in prioritising the activities.

A clean work area is pleasant for working, and work efficiency and quality shall improve. An unclean and congested area reduces the work efficiency. When the working environment is clean and peaceful, man can think and find solutions to problems.

Clean and orderly maintained work area

Seiton means systematic arrangement of the materials. We need to arrange these in such a way that we get them when needed. No one should waste time or energy for searching these materials. Arrange the processes in such a sequence that the operation costs shall be least with the assured results. This includes the re-layout of processes and bypass systems and also organisation structure. Systematic arrangement includes the fixation of place for each item, colour codification and signboards for identification and display of standard work practices and instructions. This shall reduce the searching time and reduces stress. The excess inventory shall be visible which may help in controlling and finally may reduce the cost of inventory.

It is very difficult to locate a person or a department in large offices. Such buildings and offices should display maps of the floor plans at all crucial points. On the maps, the activities such as machines under repair or maintenance, machines stopped due to various reasons, etc., are marked with suitable tags or symbols. These are called 5-S maps.

Seiton concept can be used for work management also. As explained in time management, identify and categorize the work as important, not important, urgent and not urgent and then make a schedule. Work as per the schedule. By this your works shall be more effective.

Seiso talks of spic and span cleaning. This concept includes the cleaning of everything at work place and inspection for abnormalities during cleaning and routine maintenance, thus reducing the chances of breakdowns.

If the working area is clean, the defects shall be visible. If the system is clean, the deviations shall be visible. Design the process control systems so as to get the optimum efficiency with assured quality. The inspection and testing suggested should be able to help the people to take the corrective action. There is no meaning of getting test results, where we do not know how to control. Standardize the processes to get consistency in quality and productivity. Remember that success lies in consistency and not in getting some record results. This includes the visual display of standard operating systems maintained and monitored by regular audits, which in turn would result into task accomplishment.

There should be a system to measure the improvements in various activities and areas. 5-S audit is a measurement system to measure the improvements achieved after implementing 5-S. Each time 5-S audits are carried out, these are done with a raised level of standard in order to achieve higher degree of improvement. Do not audit in the same style when you audited the system first time. As people grow, the targets should be increased. Step-by-step targets of 5-S levels can be determined in terms of audit scores based on predetermined criteria. Similarly, determine targets in each audit. The criteria changes from time to time depending on the improvements taking place in the organization, society, level of living, technical developments, etc.

Monitor the process on a regular basis, and provide education and training to the people involved in the operations so as to ensure uniform working. In the absence of suitable discipline, any system shall fail. This emphasises on the inculcation of self-discipline by top management, and imparting of tactical and strategic training. This shall improve the morale among the employees.

Normally when we refer to 5-S concepts, people think of only housekeeping, and do not think that the same can be applied for management processes like finance management, process management and work management also. One needs to think and apply his mind to find out the best methods for doing his work and improve work quality.

1.8   Organization structure and empowerment

One might think logically and develop appropriate methods for doing the work and achieving the results, but it shall not be possible to do it unless there is clarity in the organization about the authorities, responsibilities and empowerment for doing the work. The work quality depends on the clarity in the organization about the roles and responsibilities of each individual. The organization chart should be clear and explain the areas of operations of each individual. The organization charts are prepared to have clarity as to how we organize our activities, to have clarity of the responsibilities one should share and to have clarity as to where we should approach.

1.8.1   Importance of organization chart

The organization chart gives clear indication of the area of people working, who is accountable for what, who is reporting to whom, and the inter relation between posts. It is the foundation for an organization to work efficiently.

When organization structure is not clear, people occupy the places as per their convenience.

In textile mills, especially where employee attrition is high, the management tries to attract people by providing higher remunerations added to decorative designations. The designations like vice president, senior vice president, president, general manager, deputy general manager, senior general manager, additional general manager, etc., have become very common, whereas three decades back, there used to be only one person with the designation of 'manager' for a textile mill. The senior most functional designation was master, like carding master, spinning master, sizing master, bleaching master, dyeing master and folding master. The trend of giving decorative designations started creating more confusion. A person joins a mill because he was given the designation of general manager, whereas he is asked to report to a vice president, above him there is a senior vice president and then an additional president, the president who is reporting to CEO. The CEO is a puppet of some director or relation of managing director, who is not empowering any one to work freely using their knowledge and experience. The newly joined general manager feels that he was happy as a junior weaving master in his old company where he had sufficient empowerment to take decision and act so that he could show efficiency. He left that company because of the high salary and the designation offered, but after seeing the system here he is disappointed and started looking for another job.

The workers and staff approach the senior person who brought them to the mills irrespective of the designation the person has. In an organization where the organization structure is not clear, the posts are created to please a new person and some area is allotted to him, and when that man leaves, the work is allotted to another as additional responsibility, without providing any training or support. People get frustrated in such situations. This is one of the main reasons for very high attrition of technical staff in the textile and garment industry.

Another problem normally seen in textile industry is that the newly appointed person is given more remuneration and higher perks compared to the people who are working in the same company from a long time and knowing the systems and giving the results. The new person takes 3–6 months to adjust self to the new environment. Such people normally fail as they do not get the required cooperation and support from the people working from long with a low salary. It is normal that whenever a new person joins on a senior post, he brings his own people and removes the earlier people with one reason or the other. However, the new team also fails as they cannot adjust themselves to an environment which does not have proper organization structure and clear authorities, responsibilities and empowerment. New team starts the work with good enthusiasm, but when they realize that they are not empowered, and some one behind the screen is controlling, they get fed up and leave the job.

In number of family-managed mills and also in some of the so-called professionally managed companies, the key posts are occupied by the relatives of the owner or a close friend of CEO, with an intention that money is handled properly and misuse is not made. However, it is seen that these people, who are close to top man, make money for themselves, and the top man sees it helplessly as he is a close relative. They insist on purchasing the machines and spares, raw materials, colours and chemicals and other materials from their close agents without verifying the quality and price. The technicians are forced to produce quality and productivity, and they are branded as inefficient. The frustrated good technicians leave the company.

In government industries and in cooperative mills, we can see people are employed not on the basis of their competence but by the group or caste to which they belong to. These employees are vote bank to the politicians of that area. They are not having any inclination to work, but insist on giving attendance to them and salary on regular basis. The system of reserving jobs to certain category of people was started with good intention; but it is killing the creativity of the people which they had earlier. Unless a competitive environment is created, the creativity of the employees will not surface out. Such employees are forgetting that they have ability, as they are assured of promotions even without showing any results. They are not feeling the necessity of becoming competitive in the changing environment. Such companies are getting closed down as they have become unviable, and dialogues are taking place to handover such sick units to private, but the private organizations are not ready to take over such mills because of the poor work culture developed and protected. By providing reservation the work quality never improves. By working in a protected environment or reservation system one cannot win in the competition, may be in industry or in sports. Unless one works and gains experience and becomes strong because of this knowledge and experience, skill and dedication, he cannot win in the competition. The team employing such people also shall collapse. One has to exercise to keep his body fit.

1.8.2   Structure of organizations

Organization is a mechanism or structure that helps people to work effectively together. The development of society with different forms of activities demands organization. Organizing has enabled men in groups to apply their brain and muscle power to the natural resources for the production of material groups. An organizational structure consists of activities such as task allocation, coordination and supervision, which are directed towards the achievement of organizational aims. It can also be considered as the viewing glass or perspective through which individuals see their organization and its environment.

Organization implies not only a purpose or a set of purposes but also a form appropriate to carry on activities to achieve the objectives. The greater the range of activities, the structure becomes more complicated and specialized. As business grows, the separation and the definition of functions increase, but the oneness of the whole is important. With the growth of business, the departmentalization and differentiation of structure and functions becomes the need of the organization.

An organization can be structured in many different ways, depending on their objectives. The structure of an organization will determine the modes in which it operates and performs.

1.8.3   Prerequisites to achieve harmony in organization

Harmony in an organization is very important even for its survival, forget about achieving the targets. The prerequisites to achieve harmony include an orderly structural plan, definite and balanced departmentalization by major functions, the provision of means for coordination of purposes, policies and activities at every level of company and channels for communication and participation. Organization is the process of identifying and grouping the work to be performed, defining and delegating responsibility, and authority and establishing relationships for the purpose of enabling people to work most effectively together in accomplishing objectives.

The structural plan needs to indicate the span of controls and also the area of interaction. The span of control is indicated by arrows showing downwards, where as an interaction is indicated by a double-headed arrow. Some use a symbol of tennis ball to indicate two-way conversations.

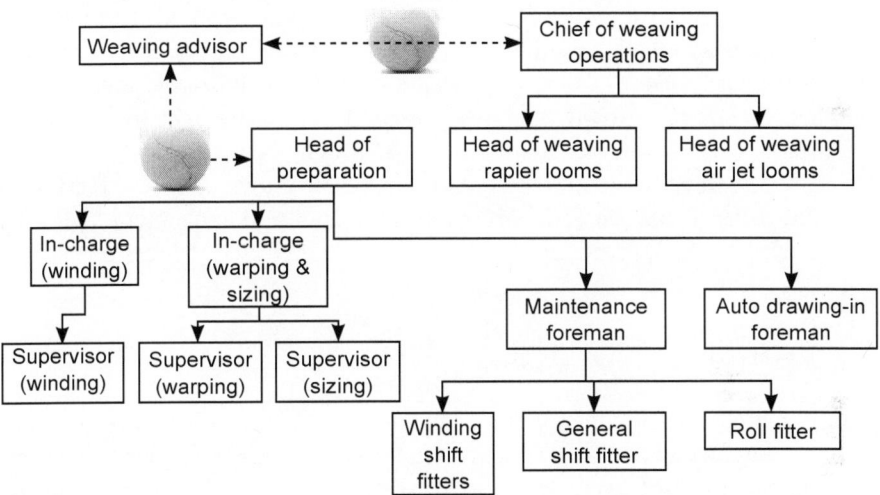

Identification of different activities and grouping them, delegation of responsibility and authority and establishment of relationships are essential for managing an organization.

Organizational structure allows the expressed allocation of responsibilities for different functions and processes to different entities such as the branch, department, workgroup and individual.

Organizational structure affects organizational action in two big ways. First, it provides the foundation on which standard operating procedures and routines rest. Second, it determines which individuals get to participate in which decision-making processes, and thus to what extent their views shape the organization's actions.

1.8.4   Organizational structure types

Organization structure depends on the type and size of organization, their mission and vision, the dependency on the technology or the skill of humans and so on. They can be grouped as pre-bureaucratic structures, bureaucratic structures, post-bureaucratic structures, functional structures, divisional structures and matrix structures.

1.8.4.1   Pre-bureaucratic structures

Pre-bureaucratic (entrepreneurial) structures lack standardization of tasks. This structure is most common in smaller organizations and is best to solve simple tasks. The structure is totally centralized. The strategic leader makes all key decisions and most communication is done by one-on-one conversations. It is particularly useful for new (entrepreneurial) business as it enables the founder to control growth and development. They are usually based on traditional domination or charismatic domination in the sense of Max Weber's tripartite classification of authority. As the industry grows, they find it difficult to changeover to bureaucratic structure of functional management styles, as the owners never appreciated the loyalty of the qualified professionals, and are always afraid that someone taking over from them.

1.8.4.2   Bureaucratic structures

A fully developed bureaucratic mechanism has an upper edge over the other ordinary organizations in the same way like machine has over the non-mechanical modes of production. Precision, speed, clarity, strict subordination, reduction of friction, and management of material and personal costs form the integral part of a strictly bureaucratic administration. Bureaucratic structures have a certain degree of standardization. They are better suited for more complex or larger scale organizations. They usually adopt a tall structure. It is very much complex and useful for hierarchical structure organization, mostly in tall organizations. The major drawback with bureaucratic structure is slow decision, because of which number of opportunities are lost. In pre-bureaucratic structure, the owner is the sole authority, who takes risk and decides fast, and hence does not miss an opportunity. In textile industry, where competition is very high, people do not like to have bureaucratic approach.

1.8.4.3   Post-bureaucratic structures

The ideas of post-bureaucratic structure developed after 1980. They include total quality management, culture management and matrix management. The decisions are based on dialogue and consensus rather than authority and command. The organization is a network rather than a hierarchy, open at the boundaries (in direct contrast to culture management). The success of the organization depends on the cohesion the team members have, and the interaction between teams as a single team. Unless the members are competent enough to understand the situation and are selfless, these structures also fail.

1.8.4.4   Functional structures

Employees within the functional divisions of an organization tend to perform a specialized set of tasks, leading to operational efficiencies within that group. Normally in large composite mills this type of organization is seen. It could lead to a lack of communication between the functional groups within an organization, making the organization slow and inflexible. Coordination and specialization of tasks are centralized in a functional structure, which makes producing a limited amount of products or services efficient and predictable. The success depends on the skill of the chief executive and making the functional heads cooperate with each other.

1.8.4.5   Divisional structures

Also called a "product structure", it groups each organizational function into a division. Each division contains all the necessary resources and functions within it. Divisions can be categorized from different points of view, like on a geographical or product/service basis. This type of structure can be seen where multiple products are produced like cotton yarn dyed shirting, cotton piece dyed bottom weights, polyester suiting, poly-wool suiting, polyester filament shirting, furnishings, terry towel, etc., in the same mill. As the work culture required is different, these are produced in different sheds and managed by different group of people. All activities including recruitments, purchases, production planning, design and development, productions, processing, packing, quality control and marketing are done separately. The chief executive and managing directors may be common. The biggest disadvantage in such structures is that the people do not develop feeling of oneness and the common functions such as engineering, finance, information technology, etc., finds difficulty in prioritizing their activities.

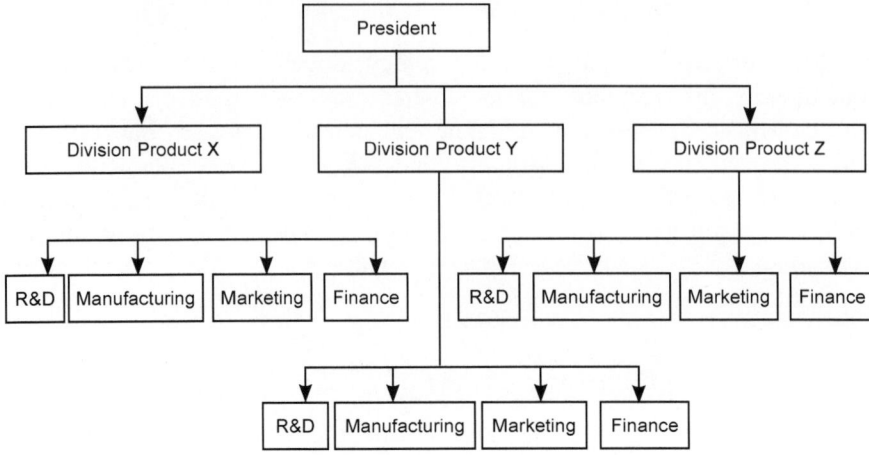

1.8.4.6   Matrix structures

Matrix structure categorizes employees on the basis of both function and product. It frequently uses teams of employees to accomplish work, in order to take advantage of the strengths, as well as make up for the weaknesses, of functional and decentralized forms. This type of structures can be seen in garment industries.

1.8.5   Span of control

The extent of controls exercised is termed as span of control. It should be ensured that one person reports to one boss only and not to multiple bosses. Multiple bosses

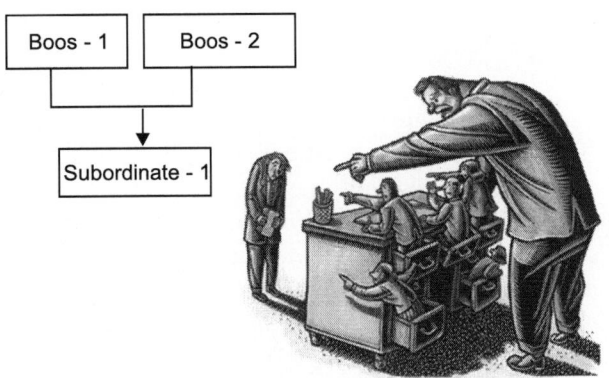

The number of people directly reporting to one boss also should be controlled. The ideal is three or four functions reporting to one boss and not more than that. A multiple reporting as shown in the chart cannot give fruitful results as the boss cannot concentrate on many and people shall be waiting for decisions. This is a normal problem in majority of family-owned textile mills, where the company chief has no confidence or trust on the senior officials working for him, and does not want anything to happen without his permission. The root of the problem is employing his relatives and friends in key positions just because they are related, who have no education or formal training required to manage those functions. In a small organization it might work, but as the organization grows, it will not work. The people get fed up and leave the organization.

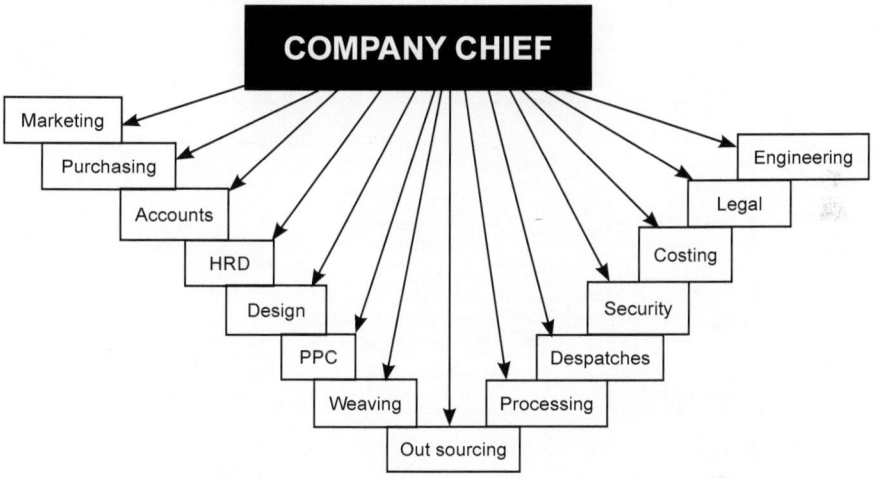

The reporting can be direct or indirect. The indirect reporting is shown in dotted lines.

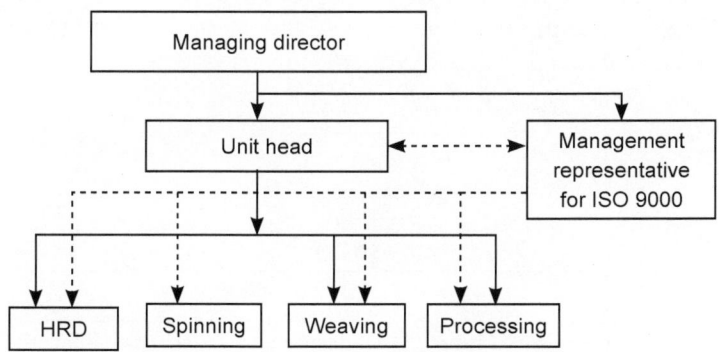

1.8.5.1 Line organization

In line organization the command goes from top to bottom in a structured hierarchy. This is also referred as military structure.

There shall be no interaction with people in the same level. They need to refer to the top authority of their section. Reporting goes from bottom to top as per the hierarchy. The decision is taken only by the top executive.

Line organization structure operates at shop floor level where one type of activity is done. In a textile mill we can see line organization in production areas like spinning, winding, weaving, etc.

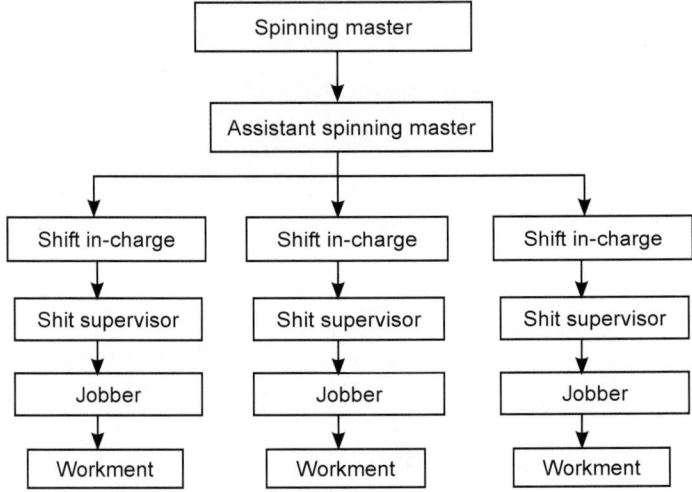

1.8.5.2 Functional organization

In a functional organization, the heads of each function shall be specialists in their work, and freedom is given to them in selecting the parameters, adjusting the manpower, selecting the raw materials and so on; however they need to operate within the sanctioned budget by the top management and are expected to show the results as agreed.

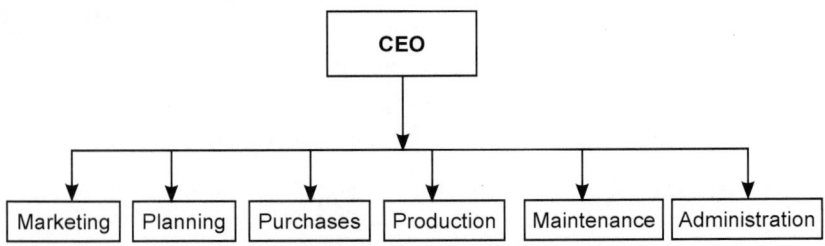

In a functional organization all functional heads report to CEO. Functional heads have authority to interact with another functional head relating to the inter connecting activities. The people down the line have no authority to interact with the people of another function without the permission of their functional heads. Functional heads have freedom to take decisions within their function.

Functional organization structure is like an inverted tree with number of branches and all are connected to one stem.

1.8.5.3   Flat structure

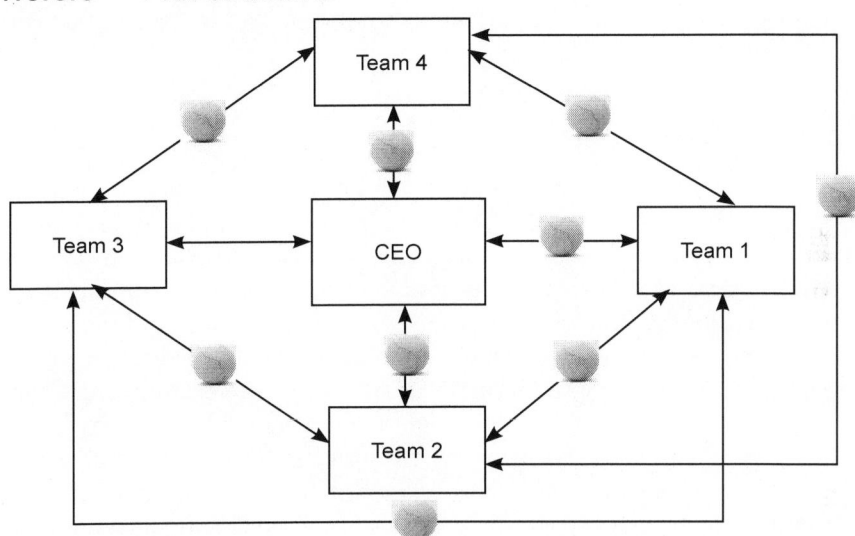

In a flat organization all teams are on same level, and each team is empowered to interact with other teams relating to their works and take decision suiting to both. CEO shall not interfere in the team decisions, but only verifies its alignment with the company policies, objectives and goals and the legal and regulatory requirements. Flat organizations operate at corporate level having number of units at different locations and having different activities. The flat structure is common in entrepreneurial start-ups, university spin offs or small companies in general. In textile and garment marketing chains the flat organizations are very effective; each team working in a different segment with different specialized product. How each team cooperates with others in taking out the unsold materials and selling at different place without competing among each other is very important. Quality of work at marketing can improve by mutual trust and cooperation and not by competition within the team members or within team.

- Each section and individuals are supposed to carry out certain activities.
- One needs to understand the purpose of the activity and the goals to be achieved.
- Design the activity to achieve the goal at the lowest possible expenses.
- Explain clearly the roles and responsibilities and works to be done by each one in the team.
- Monitor the activities.

The success of a flat organization and the quality work depends on how people work together as a team and support other teams in their works. The feeling of oneness is more important. The personal ego is the worst enemy that can fail the organization.

1.8.5.4   Network structure

The new network organizations contract out any business function that can be done better or more cheaply. Managers in network structures spend most of their time in coordinating and controlling external relations, usually by electronic means. The tasks, such as designing, getting new designs approved from the customer, collecting market information, preparing forecasts of market trends, collecting information on competitors and helping management in making strategic decisions, normally adopt a network structure. There shall be no restriction of reporting, timing and the targets for each day as in line organization. The quality of work depends on the freedom to mingle with people and come out with new ideas for the benefit of organization. This type of organization is popular in leading brands dealing with garment marketing.

1.8.5.5   Virtual structure

A virtual structure is a boundary-less organization. The virtual organization does not physically exist but is enabled by software to exist. The virtual organization exists within a network of alliances, using the Internet. The members are doing their work independently but join for doing a particular task over the Internet and exchange their views. It may be your friend circle that you contact to solve a problem of yours, may be a website to get a name for your new brand, the online marketing services on web or any such thing which has no boundary or a fixed team of people working together. The quality of work depends on the clarity in the messages given, and sincerity in replying to the messages received. This type of structure can be used for design and development for garments and fabrics.

1.8.6 Describe jobs

After deciding organization structure, decide the jobs allocated for each person. Decide the responsibilities of each person. The same job should not be given to two different designations. Same responsibilities should not be given to two people. Each person should report to only one person. Then only quality of work can be achieved in a company.

If the job descriptions are clear with the descriptions of the works to be done, the authorities and responsibilities, along with the work procedures and work instructions, the employees shall be free and empowered to do their jobs. Their enthusiasm increases when they are clear about their role in the organization. The major problem in textile and garment industry is that the jobs are not specified clearly, the authorities and responsibilities are not clear, and there is always a fear of failure haunting the enthusiastic people. In number of companies with ISO 9001 certificate, the job descriptions are written but are remaining only on papers. The ISO auditors audit the system against the work procedures written and not cross check with the job descriptions.

1.8.7 Freeze designations

A good company freezes designations, and person while joining shall be clear about his designation, the work to be done, the persons working under him and the person to whom he should report, his authorities and responsibilities so that there would be no confusion. Creating a new designation is a negative factor, although it helps in motivating one person, but demotivates number of others. Each designation should have a sanctity and respect. Just by giving decorative designations, the designations lose their value. For example, the designation of spinning master was of a very high value, but now it has lost its value. Now a spinning master is known as spinning manager, general manager spinning, president (spinning), and so on. The term 'executive' was a highly respected designation once, but now a fresh graduate is given the designation of executive, and he does the work of a junior supervisor or overseer working in earlier days. Even the designation of managing director has lost its value in number of cooperative spinning mills, who does not have powers more than that of an earlier spinning master. The CEOs are restricted only to a small section of a textile mill and the sales, purchase, human relations, projects, finance, costing, etc., are controlled by either managing director or some relatives of the managing director. Such designations after losing their values become a negative motivating factor.

1.9   Adhering to safety regulations

Textile industry is labour and skill oriented, and the skilled human is the wealth of a company. One can invest huge amount and buy latest machines, but without suitable person to man it, it becomes a waste. The success of a company depends on the efficiency of workers. There is a close relationship between safety measures and the efficiency of workers. Efficiency results in increasing the average output per worker. It is reflected in increased productivity. Safety measures are concerned not only with the physical efficiency, and safety of the workers, but also with his general well-being. Lack of safety exposes workers to health hazards. It also involves occupational health risks. People would not like to continue working in a company where safety systems are not respected and implemented, as all are concerned about their safety and well-being of their family members.

1.9.1   Safety measures and efficiency of workers

Indian workers are generally considered to be less efficient as compared to workers in other countries. Such a statement does not reflect any inherent deficiency on the part of workers. It is due to longer hours of work, low wages, poor living conditions and poor health and safety measures provided in factories. The workers in Surat, Tarapore, Silvasa and in number of places in South Gujarat and North India work daily for 12 hours and that too without weekly off. These workers are not paid any overtime wages as per Factory Act for the extra hours they work. They do not have good place to stay. Their living conditions are very bad. Four to six persons stay in a room, prepare their food, clean their clothes and attend the work for 12 hours. Their families are at their native. Climatic factors, illiteracy, low standard of living, and stating without family may also affect the efficiency adversely, but the poor working conditions happen to be the main reasons. Working environment in the factory is not conducive to increased efficiency of worker. Under unhealthy surroundings, we cannot expect workers to put in hard and sustained work. Safety measures as listed in Factory Act partly prevent workers from being exposed to the risk of accidents, and protection against dust and fumes and inflammable gases. These measures are partly welfare in nature, e.g. preventing employment of young person on dangerous machines. Other safety measures reduce the strain from working under difficult conditions.

It is necessary for all in the industry to plan their activities in such a way that it is safe: follow the safety regulations, provide safety gadgets, implement safe systems, periodically audit the safety systems and correct the lapses, educate others on safety issues and achieve targets without mishaps, without physical loss to the people working for achieving the targets.

The factory inspectors in the so-called industrially developed state of Gujarat are not doing their duty of implementing Factory Act and safety measures in the industry. Even the minimum safety regulations are not being followed leading to number of accidents, which are not being recorded. The managements are happy with the government as they are not insisting implementation of Factory Act. This is creating a very poor work environment and the people working in the industry are not happy. They are just doing the work as they do not have alternatives. The moment they get an alternate safe work environment, shall be leaving the company. The employee attrition is very high indicating that the work quality is not good. The industry owners are not realizing the fact and are continuing with the same system which they are following since long.

1.9.2   Normal causes of accidents in a textile mill

Accidents are common in poorly maintained textile mills. The accidents on work can be grouped into four main categories: accidents due to worker's mistakes while working on machines, accidents due to machine malfunctioning, accidents in material handling and accidents due to poor housekeeping.

The worker's mistakes while working on machines includes wearing loose clothes, coming drunk to the work, inadequate training, bypassing of safety switches, deactivating safety locks and not following the safety precautions, using compressed air for cleaning themselves and so on. A loose cloth can be caught in the machine while worker is concentrating on some other point in work such as creeling a bobbin on the top of the machine. A drunken worker cannot concentrate on work and can touch a running part or fall on a running machine resulting in an accident. A worker, if not trained well can commit mistakes and get himself injured. Using compressed air for cleaning the clothes and hair after the work looks very easy and majority of the workers including senior managers opt for it. Air coming at high pressure can puncture delicate parts of skin. Sometimes we get fine metal parts, water particles, etc., along with the air. The bypassing of a safety switch done for maintenance and checking purposes are not reverted in number of cases, resulting in major accidents.

| Lift without a door can lead to major accidents | Haphazard way of keeping materials lead to accidents |

Accidents due to machine malfunctioning can be reduced by improving the maintenance. Improper maintenance and settings, poor humidity conditions leading to lapping, not calibrating the pressure gauges, not monitoring the pressure vessels and compressors are some of the causes for accidents.

Accidents in material handling include overloading the trolleys, obstacles in the alleys, narrow alleys, not maintaining the material handling equipments, not closing or locking the trolleys while in movement, slippery pathway in which trolleys have to move, insufficient light in passages, lifts without doors, chains with weak links in the cranes and hoists and so on. When the height of the materials loaded is high the worker shall not able to see the passage, he may hit at some place. In number of cases it is seen that the material carrying trolleys are not provided with brakes, and worker cannot control the speed.

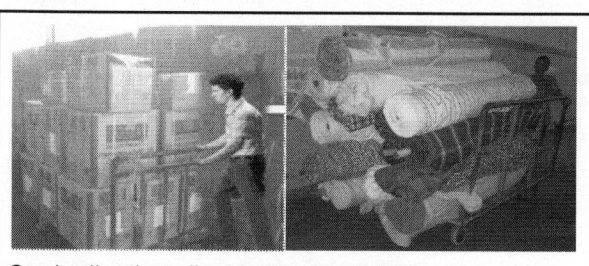

| Overloading the trolley blocking the visibility | Kicking the empty warp beams |

In warping and sizing units, the workers have a habit of kicking the empty beams for taking them to the required place. Sometimes the beams rolls fast and hit someone coming from other side.

Accidents due to poor housekeeping include blocking the entrances, piling up of unnecessary materials near electrical control boards; keeping clothes, bags, tiffin boxes on the machines or on control panels; keeping materials in passages, and so on. These are all indicators of poor work culture.

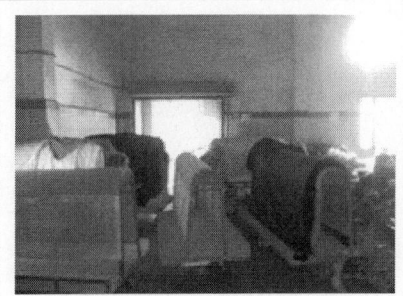

Blocking the passage by trolleys. No place for free movement.

Poor housekeeping

1.9.3   Environment, hygiene and OSHAS norms

The textile industry consists of a number of units engaged in spinning, weaving, dyeing, printing, finishing and a number of other processes that are required to convert fibre into a finished fabric or garment. There are several safety and health issues associated with the textile industry. The major safety and health issues in the textile industry are exposure to cotton dust, exposure to chemicals, exposure to noise and ergonomic issues.

1.9.3.1   Exposure to cotton dust

When cotton is processed, it emits fine cotton dust particles into the air. These particles are breathed into the lungs by the person working with the fibre. Sometimes the person can have an allergic reaction which is similar to an asthma attack. This allergic reaction causes the small airways in the lungs to contract so that air cannot quickly leave the lungs. Any air that is already in the lungs at the time of the attack has to force its way out of the body through narrowed lung passages which in turn produces wheezing sounds that are common during asthma attacks. Even if a person working with cotton does not display any allergic reactions, there is scientific evidence that people who are exposed to cotton dust may develop a permanent decrease in their breathing ability. This cotton-dust-related disease is known as Brown Lung or byssinosis, and affects thousands of people in the textile industry who are exposed to large quantity of dust. The symptoms of this disease include tightening of the chest, coughing, wheezing and shortness of breath.

The workers engaged in the processing and spinning of cotton are exposed not only to significant amounts of cotton dust but also to particles of pesticides and soil. Exposure to cotton dust and other particles leads to respiratory disorders. The Occupational Safety and Health Administration (OSHA) made it compulsory for employers in the textile industry to protect their workers from over exposure to cotton dust and its evil effects. The OSHA determined certain guidelines which can be used by all. OSHA has laid down a Cotton Dust Standard with a view to reducing the exposure of the workers to cotton dust and protecting them from the risk of byssinosis. It has set up permissible exposure limits (PELs) for cotton dust for different operations in the textile industry. This standard has helped in bringing down the rate of occurrence of byssinosis significantly. Different states might adopt different standards for occupational safety and health; however, in those states where there are no standards fixed by the State, the OSHA Cotton standards are accepted.

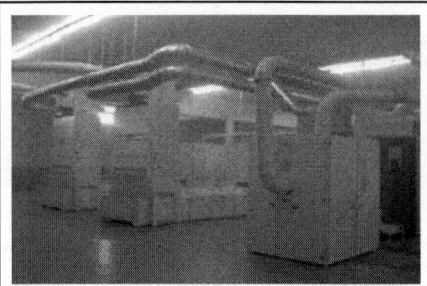

Good blow room line where all dusty air is taken out and filtered. In the department, there is no dust in spite of cotton being opened

The line without fool proof dust evacuation. The driving chain has no guard, which may lead to accidents.

OSHA has given dust standards for various operations of a textile mill. Employers are required to measure the quantity of respirable cotton dust once in 6 months or whenever there is any change that might lead to a change in the level of dust. If the level of dust in the atmosphere is higher than that as per OSHA guidelines, the management should take measures to reduce the same. As per these guidelines, the employer is required to inform the employees in writing of the dust level present in the atmosphere as well as the steps that the management is planning to take for its reduction. If the dust level cannot be reduced, it is the duty of the management to provide respirators to the employees.

Dust in textile mills not only affects the lungs, but also affects the eyes. A study conducted by a government factory inspector's report in Great Britain during 1950, indicated that out of 247 cases studies, 120 was due to flying particles hitting the eyes. With the present system of micro dust monitoring and return air ventilation systems, the flying objects have greatly reduced.

	Using mask in the areas where there are chances of dust liberation is very essential to protect people from getting lung infections.
Wearing mask while at work. The mask should cover both nose and mouth.	

1.9.3.2   Exposure to chemicals

In addition to lung problems caused by dust, textile workers who work with dyes or chemicals can develop skin allergies or rashes known as dermatitis. Finishing agents such as formaldehyde used in permanent press materials can cause allergic reactions that affect the respiratory system. Also, textile workers who are regular smokers working with dusts or finishing agents are at a higher risk of developing lung and heart problems. The risk multiplies with the amount of exposure. The duty of the employer becomes even greater in making sure that workers are not exposed to large quantities of foreign substances such as cotton dust or chemicals.

Workers in the textile industry, engaged in the activities of dyeing, printing and finishing, are exposed to a number of chemicals. Chemicals based on benzidine, optical brighteners, solvents and fixatives, crease-resistance agents releasing formaldehyde, flame retardants that include organophosphorus and organobromine compounds and antimicrobial agents are used in textile operations. Studies have revealed links between exposure to formaldehyde and nasal and lung cancer as well as to brain cancer and leukemia, which can be fatal. In the long run, exposure to formaldehyde could lead to respiratory difficulty and eczema. Contact of the chemicals with skin as well as inhalation of the chemicals can lead to several serious health effects.

1.9.3.3   Exposure to noise

High levels of noise have been observed in most of the units engaged in the textile manufacturing. In the long run, exposure to high noise levels has been known to damage the eardrum and cause hearing loss. Other problems like fatigue, absenteeism, annoyance, anxiety, reduction in efficiency, changes in pulse rate and blood pressure as well as sleep disorders have also been noted on account of continuous exposure to noise. Lack of maintenance of machinery is one of the major reasons behind the noise pollution. Though it causes serious health effects, exposure to noise is often ignored by textile units because its effects are not immediately visible and there is an absence of pain.

1.9.3.4   Ergonomic issues

Ergonomic issues are observed in a majority of the units engaged in textile-related activities in India. Most of these units have a working environment that is unsafe and unhealthy for the workers. Workers in these units face a number of problems such as unsuitable furniture, improper ventilation and lighting, and lack of efficient safety measures in case of emergencies. The workers in such units are at risk for developing various occupational diseases. Musculoskeletal disorders like carpal tunnel syndrome, forearm tendinitis, bicapital tendinitis, lower back pain, epicondylitis, neck pain, shoulder pain, and osteoarthritis of the knees are some of the occupational diseases that have been observed among the workers on account of poor ergonomic conditions. There is a considerable difference in the heights of the stools and the tables used for various operations such as cutting and ironing in garment industries. This makes the workers to sit in an uncomfortable position for entire work days. The stools are not padded in most of the units, leading to increased discomfort on the part of the workers. Number of stools does not have a backrest, as a result of which the workers do not get adequate support to the back. In most of the units, the level of lighting is low and improper placement of lighting fixtures lead to low lighting at the point of work, leading to eye strain. On account of the continuous use of irons in some units, the humidity level is very high, contributing to the workers discomfort. Apart from this, lack of efficient measures for the safety of the workers is also observed, may be due to negligence of the management or due to lack of awareness among workers. Lack of essential items such as first-aid kits, fire extinguishers, and alarms can be seen in most of the units. This puts the workers under great risk in times of an emergency. Protective equipments like metallic gloves were not provided to the workers in several units for protection against potential accidents and injuries.

It is essential that the workers be aware of the various occupational hazards in the industry and use safety gadgets. At the same time, it is necessary that the management take the necessary steps to protect workers from potential hazardous situations. The following points are necessary to improve the safety and health conditions in textile units:

1. The seats and the table's alignment in height to prevent musculoskeletal strain.
2. Proper lighting at the place of work to prevent strain on eyes.
3. Maintenance of machinery to reduce the level of noise.
4. Providing ear plugs to reduce exposure to noise.
5. Job rotation to avoid continuous noise exposure for a long period of time.
6. Proper ventilation at the place of work.
7. Providing masks to reduce the exposure to dust.
8. Availability of trained medical personnel and first-aid facilities as well as safety equipments such as fire extinguishers and fire alarms at the place of work.
9. Providing safety gloves to prevent heavy exposure to dangerous chemicals and electrical shocks.
10. Setting up and maintenance of dust control equipment.
11. Conducting periodic medical examinations for the workers and taking appropriate measures in case significant occupational health problems are observed.

The passing of the Occupational Health and Safety Act (OSHA) of 1970 was the means by which industries were forced to adopt standards to ensure a safe and healthy environment for workers. OSHA guarantees to employees the right to a safe and healthy working environment, and OSHA inspectors can walk in and inspect workplaces at any time. Workers who feel that their workplace is unsafe can file complaints with OSHA. If upon inspection, OSHA finds violations of industry standards, then the industry can be heavily fined. In addition, OSHA requires that industries keep records of work-related accidents, illnesses, injuries, and exposures to harmful materials. This information must be made available to employees and government agencies and provided upon request.

1.9.4   Safety regulations

Accidents affect not only workers losing their livelihood but also employers in terms of compensation to be paid to the workers. Accidents are a significant cause of dispute between workers and management. The Factories Act 1948 has laid down certain measures for the safety of workers employed in the factories. There are numbers of safety regulations adopted by various countries and standard bodies. The safety or the people working on the machine have been given prime importance.

The safety regulations deal with number of issues. While discussing on machine lay out, it specifies minimum distance between two machines, minimum distance between wall and the machine, minimum number of entry and exit points in a production hall depending on its size, the minimum height of the roof, etc., so that the workers can move freely, escape in case of fire accidents and avoid hitting himself to a machine part or to a wall.

The safety regulations also address the preventive actions in case of fire and the type of extinguishers to be used. While addressing the first-aid boxes and safety gadgets, the safety regulations stress on the materials to be kept in a first-aid box, maintenance of first-aid box, verifying the expiry dates of medicines, providing rest room in case of accidents or illness and providing ambulance facilities.

Disposing of wastes has become a burning issue as the wastes pollute. Some of the wastes are hazardous. The mill has to identify the hazardous wastes and dispose them in a safe manner. The wastes cannot be just thrown out or dumped in a ditch. The pollution control boards have certified some agencies as authorised waste disposers, and the mills should send the wastes to only such agents. The chemical carboys need to be washed thoroughly before disposing them. The hard wastes used for cleaning machine parts, diesel engines are to be given only to authorised waste collectors.

1.9.4.1   Need for safety measures

Safety measures result in improving the conditions under which workers are employed and work. It not only improves their physical efficiency, but also provides protection to their life and limb. Inadequate provision of safety measures in factories may lead to increase in the number of accidents. Human failures due to carelessness, ignorance, inadequate skill, and improper supervision have also contributed to accidents, and the consequent need for safety measures. Other factors giving rise to the need for safety measures are rapid industrialization with its complexities in manufacturing process and layout, expansion or modifications in existing factories, setting up of new industries involving hazards not known earlier, lack of safety consciousness on the part of both workers and management, and inadequate realisation of the financial implications of accidents.

Safety measures which are provided in the Factories Act 1948 are considered to be bare minimum in terms of adequacy. Such measures are required to be effectively implemented. In addition to implementing safety measures provided in the Factories Act, there is also need for providing training in safety to workers, and installing safety equipment in the factories. Employers should take the initiative in providing training in safety to employees. Workers' unions should take interest in safety promotion. Periodic training courses in accident prevention can be organised. Safety should become a habit with employers and the workers alike. The following are some of the examples of safety norms that textile mills should follow as laid out by various safety regulations world over.

1.9.4.2 Examples of safety norms for textile mills

Fencing of machinery

In every factory, measures should be taken for secured fencing of machinery. Safeguards of substantial construction must be raised and constantly maintained and kept in position while the parts of machinery (they are fencing) are in motion or in use. Fencing is necessary in respect of every moving part of a prime mover, headrace and tailrace of every water-wheel and water turbine, every part of an electric generator, a motor or a rotary convertor, every part of transmission machinery and every dangerous part of any other machinery.

Machine guarding

An employer must ensure that power transmission parts are guarded so that no one or any material can come in contact with the moving parts.

Beater guards

When any machinery is equipped with a beater, such beater shall be provided with metal covers which will prevent contact with the beater. Such covers shall be provided with an interlock which will prevent the cover from being raised while the machine is in motion and prevent the operation of the machine while the cover is open.

Gear housing covers

Whenever gear housing is there, they should be properly covered. They should have interlocks so that the cover cannot be opened when the machines are working.

Casing

Every set screw, bolt or key on any revolving shaft, spindle, wheel or pinion shall be so sunk, encased or otherwise effectively guarded as to prevent danger in all machinery driven by power and installed in the factory.

In machineries like chain mercerising or clip stenters, a guard shall be installed at each end of the frame between the in-running chain and the clip opener, to prevent the worker's fingers from being caught.

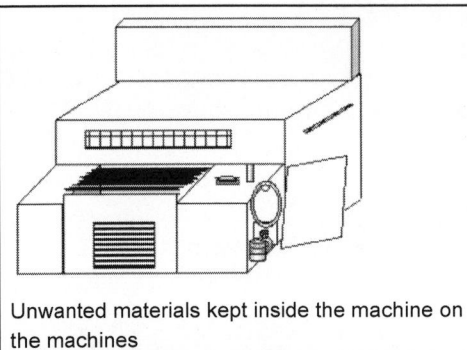

Unwanted materials kept inside the machine on the machines	Keeping materials like clothes, records, tools, wheels, etc., inside the machines should be avoided, as it is one of the main reasons for accidents. The doors should not be opened when the machines are working.

Rotary staple cutters

A guard shall be installed completely enclosing the cutters to prevent the hands of the operator from reaching the cutting zone.

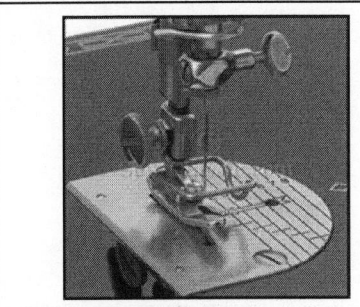

Needle guard in sewing machine	*Needle guards:* In garment factories, and also in fabric processing houses, needle guards are to be provided on all the sewing machines to protect the fingers of the tailors getting hit by needles while feeding fabric panels.

Work on or near machinery in motion

If there is a need to examine the machinery in motion, such examination shall be carried by trained worker. Such workers shall wear tight-fitting clothes and their names should be recorded in the register prescribed in this connection.

The machinery in motion with which other workers could be hurt, such machines should be securely fenced to prevent such contact. No woman or young person shall be allowed to clean, lubricate or adjust any part of a prime mover or transmission machinery, while the machinery is in motion.

Means of stopping machines

In every factory suitable striking gear or other efficient mechanical appliances shall be provided and maintained, and used to move driving belts. Such gear or appliances shall be so constructed, placed and maintained as to prevent the belt from creeping back on to the fast pulley. Driving belts when not in use shall not be allowed to rest or ride upon shaft in motion. In every factory, suitable devices for cutting off power in emergencies from running machinery shall be provided and maintained in every workroom.

With modern machines where inverter drives are provided and are controlled by computer programming, the chances of accidents are eliminated.

Every textile machine shall be provided with individual mechanical or electrical means for stopping such machines. On machines driven by belts and shafting a locking-type shifter or an equivalent positive device shall be used. On operations where injury to the operator might result if motors were to restart after power failures, provision shall be made to prevent machines from automatically restarting upon restoration of power.

Protection for the machine repairers: Whenever a machine is stopped for repairs and a person is working inside, provision should be made to prevent it from being started inadvertently. Locks may be provided to control panels or the fuses may be removed and kept out.

Push button control: Pushbutton control shall have stop and start buttons located at each end of the machines, and additional buttons located on both sides of the machine in case of long and wide machines like sizing, mercerising etc. If callender rolls are used, additional buttons shall be provided at both sides of the machine at points near the nips.

Lever control: When machines are operated by control levers, these levers shall be connected to a horizontal bar or treadle located not more than 69 inches above the floor to control the operation from any point. However, modern machines are running with direct drives, and the concept of lever control has become obsolete. However, some old mill, in some remote area may still have this type of controls. In number of textile institutes, this type of machines is maintained from the academic point of view.

Handles: Stopping and starting handles shall be designed to the proper length to prevent the worker's hand or fingers from striking against any revolving part, gear guard, or any other part of the machine.

Revolving machinery: Effective measures shall be taken in every factory to ensure that the safe working peripheral speed of every revolving vessel, cage, basket, flywheel, pulley disc or similar appliance driven by power is not exceeded. A notice indicating the maximum safe working peripheral speeds of each revolving machine shall be put up in every room in a factory in which the process of grinding is carried on.

Cleanout holes: Cleanout holes used for cleaning out the materials from a machine, mainly in blow room, within reaching distance of the fan or picker beater shall have their covers securely fastened and they shall not be opened while the machine is in motion. A sign board is to be put indicating it as danger and not to put hands when the machine is in operation.

Cleanout hole Secured with screws

Nip guards: The feed rolls, wherever provided shall be covered with a guard designed to prevent the operator from reaching the nip while the machinery is in operation. All nip guards shall not be more than the specified gaps shown in Table the table on next page (Refer the figure to understand better).

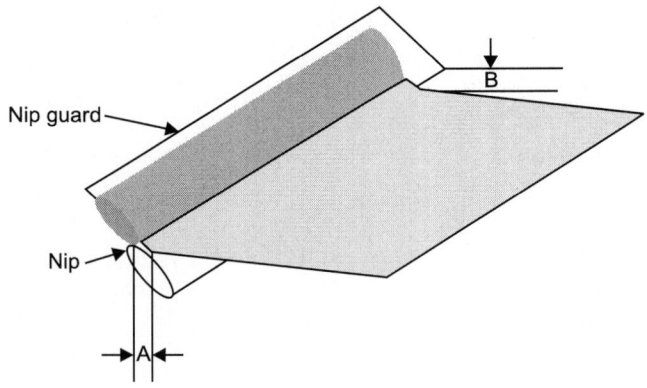

Distance of opening from nip point in inches A	Maximum width of opening in inches B
0 to 1.5	0.25
1.5 to 2.5	0.375
2.5 to 3.5	0.5
3.5 to 5.5	0.625
5.5 to 6.5	0.75
6.5 to 7.5	0.875
7.5 to 8.5	1.25

Feed rolls: The feed rolls on all opening and picking machinery shall be covered with a guard designed to prevent the operator from reaching the nip while the machinery is in operation.

Enclosures: In a machine having wire points, strikers or combs, they need to be enclosed and provided with guards. The type of enclosure depends on the type of points and the speed in which they operate. For example carding machine cylinder and lickerin shall be equipped with guards and the doffers shall be enclosed. It means cylinder and lickerin should not be able to work if the doors are open. Doffer, which runs slow, can work. Similarly the combers should not work if the doors are opened.

Pin guard: A guard shall be placed ahead of the feed end of a machine using falling pins like a gill box, and shall be so designed that it will prevent the worker's fingers from being caught in the pins of the intersecting fallers.

Safety belts: Where workers have to work at a height and support is less, they need to be provided with safety belts along with safety jacket and helmets.	
Safety limit switches 	Safety limit switches: Safety limit switches are to be provided for all the doors of any textile machine, which when opened while running may be prone to cause accidents. There are varieties of limit switches available, which need to be selected depending on the type of the machine. They need to be maintained and monitored.

Pressure plant: If in any factory, any part of the plant or machinery used in a manufacturing process is operated at a pressure above atmospheric pressure, effective measures shall be taken to ensure that the safe working pressure of such part is not exceeded. These machines are to be inspected by a competent authority on a periodic basis, and to be certified as fit for use.

Hoists and lifts: In every factory hoists and lifts shall be of good mechanical construction, sound material and of adequate strength. They shall be properly maintained, and shall be thoroughly examined by a competent person at least once in every six months. A register shall be kept containing the prescribed particulars of each such examination. Every hoist way and lift way shall be sufficiently protected by an enclosure fitted with gates, and the hoist or lift and every such enclosure shall be so constructed as to prevent any person or thing from being trapped between any part of the hoist or lift and any fixed structure or moving part. The maximum safe working load shall be plainly marked on every hoist or lift, and no load greater than such load shall be carried thereon. The cage of every hoist or lift used for carrying persons shall be fitted with a gate on each side from which access is afforded to a landing. Every gate shall be fitted with interlocking or other efficient device to secure that the gate cannot be opened except when the cage is at the landing and that the cage cannot be moved unless the gate is closed. Where in the hoists and lifts used for carrying persons, the cage is supported by rope or chain, there shall be at least two ropes or chains separately connected with the cage and balance weight, and each rope or chain with its attachments shall be capable of carrying the whole weight of the cage together with its maximum load. Efficient devices shall be provided and maintained capable of supporting the cage together with its maximum load in the event of breakage of the rope, chain or attachments. An efficient automatic device shall be provided and maintained to prevent the cage from overrunning.

Lifting machines, chains, ropes and lifting tackles: 'Lifting machine' means any crane, crab, winch, pulley block, gin wheel, and runway. 'Lifting tackle' means chain slings, rope slings, hooks, shackles and swivels. In every factory, following safety measures shall be adopted in respect of every lifting machine (other than a hoist and lift) and every chain, rope and lifting tackle for the purpose of raising or lowering persons, goods or materials —

    a. All parts including the working gear of every lifting machine and every chain, rope or lifting tackle shall be of good construction, sound material and adequate strength and free from defect, properly maintained and thoroughly examined by a competent person at least once in every period of twelve months.

b. No lifting machine and no chain, rope or lifting tackle shall be loaded beyond the safe working load which shall be plainly marked on it.

c. While any person is employed or working on or near the wheel track of a travelling crane in any place where he would be liable to be struck by the crane, effective measures shall be taken to ensure that the crane does not approach within twenty feet of that place. A lifting machine or a chain, rope or lifting tackle shall be thoroughly examined in order to arrive at a reliable conclusion as to its safety.

The person working with cranes shall be provided with helmet and ensured that he uses it.

	Protective helmet can save head from getting injured while working in hoists		Slipping from stairs is one of the normal accidents we see in industry

Floors, stairs, and means of access: In every factory all floors, steps, stairs and passages shall be of sound construction and properly maintained, and where it is necessary to ensure safety, steps, stairs, and passages shall be provided with substantial hand rails. There shall, so far as is reasonably practicable, be provided and maintained safe means of access to every place at which any person is at any time required to work.

Proper space is not provided for the forklift to turn, resulting in an accident. No railing provided at the edge of the ramp.	No railing provided and materials are stacked dangerously at the extreme.

Dangerously working on a lift without any doors or side walls.

Pits, sumps, openings in floor etc., which may be a source of danger, shall be either securely covered or securely fenced. Securely fencing a pit means covering or fencing it in such a way that it ceases to be a source of danger.

	*Excessive weights:* No person shall be employed in any factory to lift, carry or move any load so heavy as to be likely to cause him an injury.
Securing ladders: The ladders, wherever used should be secured or someone should be holding it tight. Slipping from ladder can lead to serious injuries. Foldable ladders with support in the bottom and broad bottom is recommended.	
Personal protective equipment: Workers engaged in handling acids or caustics in bulk, repairing pipe lines containing acids or caustics, etc., shall be provided with personal protective equipment to conform to the requirements like gum boots, hand gloves, masks and caps. Ensure that people use the personal protective equipments.	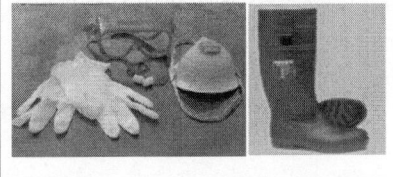
Protection of eyes: If the manufacturing process carried on in any factory is such that it involves risk of injury to the eyes from particles thrown off in the course of the process or risk to the eyes by reason of exposure to excessive lights, effective screens or suitable goggles shall be provided for the protection of persons employed on, or in the immediate nearness of, the process.	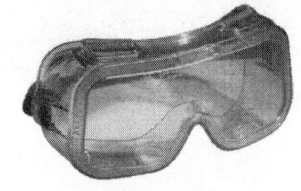 Safety goggles

Precautions against dangerous fumes and use of portable light

i. No person shall enter any chamber, tank, vat, pit, pipe or other confined space in a factory in which dangerous fumes are likely to be present to such an extent as to cause risk of persons being overcome thereby;

ii. No portable electric light of voltage exceeding twenty four volts shall be permitted in any factory for use inside any confined space. Where the fumes present are likely to be inflammable no lamp or light, other than of flame–proof nature, shall be allowed to be used.

iii. No person in any factory shall be allowed to enter any confined space, until all practicable measures have been taken to reverse any fumes which may be present and to prevent any ingress of fumes.

iv. Suitable breathing apparatus, reviving apparatus and belts and ropes shall be kept in every factory for instant use. All such apparatus shall be periodically examined and certified by a competent person to be fit for use.

v. No person shall be permitted to enter in any factory, any boiler, furnace, chamber, tank, pipe, or other confined space for the purpose of working or making any examination until it has been sufficiently cooled by ventilation or otherwise to be safe for persons to enter.

Explosive or inflammable dust, gas, etc.: If any manufacturing process in the factory produces dust, gas, fume, or vapour of such a nature as is likely to explode on ignition, measures shall be taken to prevent any such explosion by effective enclosure of the plant or machinery used in the process; removal or prevention of the accumulation of such dust, gas, fume or vapour; exclusion or effective enclosure of all possible source of ignition. Measures shall also be adopted to restrict the spread and effects of the explosion by providing in the plant or machinery of chokes, baffles, vents, or other effective appliances.

Rotary filters: Waste collecting units with a provision for filtering air by Rotary filters shall be provided wherever the cotton dust is being liberated in a big way like blow room and carding.

Rotary filters

Reducing valves, safety valves, and pressure gauges: Reducing valves, safety valves, and pressure gauges shall be maintained and got periodically calibrated. They shall be got audited by competent authorities periodically as per the rules prevailing in the state.

Vacuum relief valves: These are important from the safety of the company and the persons working nearby. Vacuum relief valves shall conform to the Code for Pressure Vessels, which need annual inspection and certification from a competent authority.

Drying cylinder enclosure: When enclosures or hoods are used over cylinder drying rolls, such enclosures or hoods shall be provided with an exhaust system which will effectively prevent wet air and steam from escaping into the workroom. Cylinder driers are used in sizing machines, mercerizing machines, drying range, open width soapers, etc.

Hood and exhaust on drying cylinders

Clear alley for free movement

Expansion chambers: Slasher kettles and cookers shall be provided with expansion chambers in the covers, or drains, to prevent surging over. Steam-control valves shall be so located that they can be operated without exposing the worker to moving parts, hot surfaces, or steam.

Slipping and falling is one of the most common accidents we see in textile mills

Housekeeping: Aisles and working spaces shall be kept in good order without any obstacles and space for free movement of men and materials. A congested aisles leads to number of accidents like the workers cloth or body touching the running parts of the machine, putting step on slippery materials, hitting another person coming in the opposite direction and so on.

Inspection and maintenance: All guards and other safety devices, including starting and stopping devices, shall be properly maintained. Periodic audits shall be done by people independent of the functions to ensure that all are working and performing as needed.

Lighting and illumination: Lighting and illumination shall conform to the safety and health requirements for the people working. The passages, materials stored and machines should be clearly visible, whereas the lights should not glare and harm the eyesight. Uniform and cool lighting without shadows are the requirement.

Identification of piping systems: Identification of piping systems shall conform to nation's code of identification. For example, the hydrant pipes should be painted with red colour, the gas pipeline by yellow colour.

Steam pipes: All pipes carrying steam or hot water for process or servicing machinery, when exposed to contact and located within seven feet of the floor or working platform shall be covered with a heat-insulating material, or guarded with equivalent protection.

Swiveled double-bar gates: Swiveled double-bar gates shall be installed on machines engaged in preparing beams or batches like warping machines, beam to beam rolling machine, batchers etc., operating in excess of 450 yards per minute. These gates shall be so interlocked that the machine cannot be operated until the gate is in the "closed position," except for the purpose of inching or jogging.

Closed position: "Closed position" shall mean that the top bar of the gate shall be at least 42 inches from the floor or working platform, the lower bar shall be at least 21 inches from the floor or working platform and the gate shall be located 15 inches from the vertical tangent to the beam head.

Benches or working platforms: In machines where height is more considering the height of the operator, benches or working platforms approximately 10 inches in height and 8 inches in width should be installed along the entire running length of the machine for the worker to stand on while creeling the machines. Such benches or platforms shall be covered with an abrasive or nonslip material.

Shuttle guard: On shuttle looms, each loom shall be equipped with a guard designed to minimize the danger of the shuttle flying out of the shed.

Shearing machines: All revolving blades on shearing machines shall be guarded so that the opening between the cloth surface and the bottom of the guard will not exceed three-eighths inch.

Protection for working in hot steam operated machines like J-Box, Kier etc.: Each valve controlling the flow of steam, injurious gases, or liquids into a J-box shall be equipped with a chain, lock and key, so that any worker who enters the J-box can lock the valve and retain the key in his possession.

Any other method which will prevent steam, injurious gases or liquids from entering the J-box while the worker is in it will comply with this provision.

Open-width bleaching: The nip of all in-running rolls on open-width bleaching machine rolls shall be protected with a guard to prevent the worker from being caught at the nip. The guard shall extend across the entire length of the nip.

Grey and white bins: Guard rails conforming to the general safety and health standards, shall be provided where workers are required to plait by hand from the top of the bin so as to protect the worker from falling to a lower level.

Oil cups: Oil cups shall be located to permit safe and easy access. They shall be of the extension type to permit oiling while machines are operating.
Roll arms: In jigger dyeing machines roll arms on jigs shall be built to allow for extra-large batches, and to prevent the centre bar from being forced off, causing the batch to fall.

Sanforizing and palmer machine: A safety trip rod, cable or wire center cord shall be provided across the front and back of all palmer cylinders extending the length of the face of the cylinder. It shall operate readily whether pushed or pulled. This safety trip shall be not more than 72 inches above the level on which the operator stands and shall be readily accessible.

Ironer: Each flat-work or collar ironer shall be equipped with a safety bar or other guard across the entire front of the feed or first pressure rolls, so arranged that the striking of the bar or guard by the hand of the operator or other person will stop the machine. The pressure rolls shall be covered or guarded so that the operator or other person cannot reach into the rolls without removing the guard. This may be either a vertical guard on all sides or a complete cover. If a vertical guard is used, the distance from the floor or working platform to the top of guard shall be not less than 6 feet.

Splash guard: Splash guards shall be installed on all rope washers unless the machine is so designed as to prevent the water or liquid from splashing the operator, the floor or working surface.

Centrifugal hydro-extractor: Each extractor shall be equipped with a metal cover with an interlocking device that will prevent the cover from being opened while the basket is in motion, and also prevent the power operation of the basket while the cover is open. Extractor shall be equipped with a mechanically or electrically operated brake to quickly stop the basket when the power driving the basket is shut off. Each centrifugal extractor shall be effectively secured in position on the floor or foundation so as to eliminate unnecessary vibration, and shall not be operated at a speed greater than the manufacturer's rating, which shall be stamped where easily visible in letters not less than one-quarter inch in height. The maximum allowable speed shall be given in revolutions per minute (rpm).

Safety stop bar on rope washers: A safety trip rod, cable or wire center cord shall be provided across the front and back of all rope washers extending the length of the face of the washer. It shall operate readily whether pushed or pulled. This safety trip shall be not more than 72 inches above the level on which the operator stands and shall be readily accessible.

Laundry washer tumbler or shaker: Each drying tumbler, each double cylinder shaker or clothes tumbler, and each washing machine shall be equipped with an interlock device which will prevent the power operation of the inside cylinder when the outer door on the case or shell is open, and which will also prevent the outer door on the case or shell from being opened without shutting off the power. This should not prevent the movement of the inner cylinder by means of a hand operated mechanism or an "inching device."Each enclosed barrel shall also be equipped with adequate means for holding open the doors or covers of the inner and outer cylinders or shells while it is being loaded or unloaded.

Printing machine (roller type): The engraved roller gears and the large crown wheel shall be provided with a protective disc which will enclose the nips of the in-running gears. Individual discs for each nip will be deemed to be in compliance with the provisions of Factory Act.

Callenders: The nip at the in-running side of the rolls shall be provided with a guard extending across the entire length of the nip and arranged to prevent the fingers of the workers from being pulled in between the rolls or between the guard and the rolls, and constructed so that the cloth can be fed into the rolls safely.

Hand bailing machine: An angle-iron-handle stop guard shall be installed at the right angle to the frame of the machine. The stop guard shall be so designed and so located that it will prevent the handle from traveling beyond the vertical position should the handle slip from the operator's hand when the pawl has been released from the teeth of the take-up gear.

Swing folder (overhead type): The bottom of the overhead folders shall be located not less than 7 feet from the floor or working surface.

Color-mixing room: Floors in color-mixing rooms shall be constructed to drain easily.

Dye kettles and vats: Pipes or drains of sufficient capacity to carry the contents safely away from the working area shall be installed where there are dye kettles and vats which may at any time contain hot or corrosive liquids. These shall not empty directly onto the floor.

Open tanks and vats for mixing and storage of hot or corrosive liquids: Guardrails shall be provided for open tanks and vats which conform to the requirements of factory act. A sign board shall be put for the people not to go beyond certain point.	

Shutoff valves: Boiling tanks, caustic tanks, and hot liquid containers, so located that the operator cannot see the contents from the floor or working area shall have emergency shutoff valves controlled from a point not subject to danger of splash. Valves shall conform to the Pressure Vessel Code of the factory Act for Unfired Pressure Vessels.

Acid carboys: Carboys shall be provided with inclinators, or the acid shall be withdrawn from the carboys by means of pumping without pressure in the carboy, or by means of hand operated siphons.

Handling caustic soda and caustic potash: Means shall be provided for handling and emptying caustic soda and caustic potash containers to prevent workers from coming in contact with the caustic.

Water shower: Water shower need to be provided for washing face and eyes, and workers should not wash the face by taking water in their palms as the palm might contain chemicals.	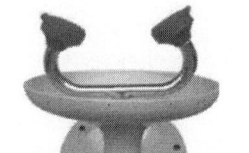

Roll bench: Cleats shall be installed on the ends of roll benches.

Workroom ventilation: In all workrooms in which potentially toxic substances are used, the maximum allowable concentrations and airborne contaminants shall be maintained as per the Factory Act. The number of air changes required depends on the type of machinery being installed. Table below gives the recommendations for spinning and weaving mills. As normally process house are not having controlled ventilation system, norms are not available.

Recommended air changes per hour

Department	Number of air changes per hour
Blow room, drawing, combing and roving	15
Carding	20
Spinning	45
Winding	30
Twisting, warping, sizing and weaving	20

Providing sufficient light: The proper amount of light under which a healthy eye should work has become a matter of importance in recent years, not only for efficiency in work but also for preventing accidents. Most authorities agree that more nearly the quality of light approaches day light the better it is. A good lighting installation should satisfy the following requisites.

1. The intensity should be ample to arable one to see clearly and distinctly.
2. The distribution of illumination should be nearly uniform.
3. The light should be soft and well diffused.
4. The source of light should be placed well above the range of vision and glare should be eliminated.

The illuminating Engineering Society of Cotton Textile Mills (England) has recommended that the actual standard for the different departments for textile industries should be as follows: opening, mixing and carding, 10 FC; slubbing, roving and spinning 20 FC; warping 20 FC; beaming 20 FC; inspection 50–100 FC; drawing in by hands 100 FC and weaving 25 FC (FC = Foot Candle).

The Factories Act prohibits employment of young persons on certain types of machines as specified under Sec.23 of the Act. They can work only after they have been fully instructed as to the dangers arising in connection with the machines and the precautions to be observed. They should have received sufficient training in work at such machines. They should be under adequate supervision by a person who has a thorough knowledge and experience of the machines.

Prohibition of employment of women and children near cotton openers: No women or child shall be employed in any part of a factory where pressing a cotton–opener is at work.

The organizations working proactively for implementation of safety measures ensure safe working conditions, and hence people can work without any hesitation and their efficiency and work quality improves. If providing safety gadgets is the responsibility of the management, using them properly is the responsibility of people working. It is seen that the staff and workers deactivate the safety systems provided along with the machine in order to keep the machines working as the productions are measured by the counters or timers fitted on the machines. They overload trolleys in order to save the number of trips which may result in dashing to someone or to an object.

| An overloaded trolley. | Ideal loading – worker can see the road where he is taking his material |

1.9.5   Precautions to prevent fires

Fire accident in textile mill is a nightmare for the industry. Numbers of well reputed mills were burnt down into ashes in fire accidents. To name a few Mysore spinning and weaving mills, Bangalore, Phoenix Mills at Mumbai, Gold Mohar Mills at Mumbai, Morarjee Mills at Mumbai and so on. Well running mills employing thousands of employees were closed down permanently.

The cotton fibre when in loose form catches fire very fast, and within no time the fire spreads. Therefore it is very essential that 100% of the staff and workers are trained for fire fighting and first aid in a textile mill and they should be educated to prevent fires.

Fires in garment factories normally result in more deaths as mainly ladies are working in a congested area, whereas in Textile mills, it is not that congested and there shall be space for people to move fast.

The fire accidents in textile mills can be grouped into four categories viz. Fires due to mechanical friction, Fires due to electrical short circuits, fires due to negligence of staff and workmen and fire due to natural reasons or nature's fury.

1.9.5.1   Fires due to mechanical friction

There are number of incidents of fire accidents in a textile mill due to mechanical friction.

In blow room, the chances of fire are very high and also each fire in blow room is a major fire because of loose cotton fibres. We often get stone pieces and metal pieces along with cotton. The stone pieces when hit the beater at a high speed, produces spark resulting in fire. This problem is more in cottons procured from old ginning mills with poor floor conditions and in blow rooms running on wastes. Providing gravity traps in blow room immediately after the bale breaker can remove the stone pieces.

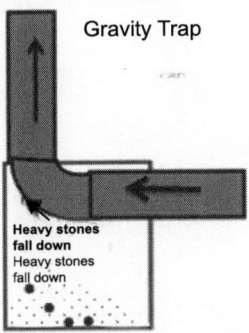

The metal pieces are another cause of fire. Providing magnets and frequent cleaning of the magnets is very important. In almost all blow rooms magnets are provided, but there is no system of cleaning the magnets frequently. In number of cases magnets are provided in the top and worker cannot reach the magnet. It is of no use. The magnet should be visible and approachable so that workers can remove accumulated cotton and metal parts. Openable windows need to be provided, which have transparent acrylic sheets so that worker can see the accumulated cotton and remove the metal piece.

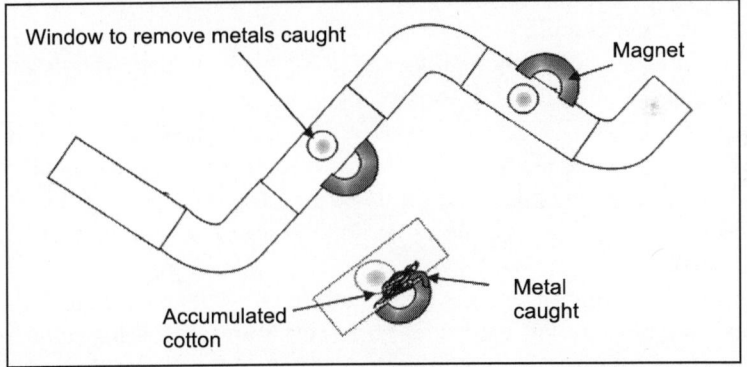

Improper suction may be due to a slack belt driving the fan or snapped belt, leads to jamming of cotton on beater which is running. This results in a fire. Periodic checking of belts is very essential. Sometimes, the improper suction may be due to leakages in the cotton conveying pipe joints.

There are cases of side leather linings wearing out in the condensers leading to jamming of condensers, which lead to fire.

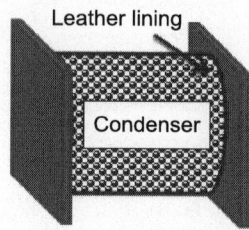

Fire in the central waste suction unit is very common because of the fine micro dust jamming the screens, and if not removed properly by the waste strippers, goes back and jams. As normally no worker shall be there to monitor, and it is not possible for human beings to control the fire when the wastes are moving in such a high speed, it is very essential to have fire detectors, fire diverting system and CO_2 flooding in blow room and waste evacuating units.

In carding, if the feeding is more and the material is not collected by doffer, there shall be loading on the cylinder. Sometimes it is seen that water drops either from humidification system or from roof. If water drops fall on cotton, that also leads to loading on cylinder. In case of surgical cotton plants, if the bleached cotton fed is not fully dry or over dry, we get loading. If loading becomes more, there shall be friction between cylinder and flats resulting in fire.

Improper settings causing wire points touching some parts like front plate or lickerin are resulting in fires sometimes.

Lapping of cotton on flat belts where flat driving pulleys are provided results in friction and high heat generation resulting in fires.

There are cases of web breaking and going back on the doffer, resulting in jam between cylinder and doffer resulting in a fire.

Lapping of fibres on the fluted rollers, especially while working viscose staple fibres on high speed draw frames, is one of the main reasons for fire in draw frame section. The lapping on fluted roller is difficult to remove. If the stop motion is working well along with the brake, this type of accidents can be prevented.

Jammed bearings generate excessive heat. This can happen in any machine like blow room, card or even in ring frames. Bearing jams are due to micro-dust entering the bearing and drying the lubricants. It is necessary to use sealed bearings wherever the chances of dust entering the bearing are there. Periodical checking of bearing is very essential.

Spindle tape entangling on tin roller pulleys and bearings in ring frames is a common reason for fires in ring frame section.

Poor quality lubricants dries the bearings fast and results in a fire. Hence one should be very strict in approving the lubricants.

1.9.5.2   Fires due to electrical short circuits

This is a common reason for fire. Loose wires, low capacity cables and wires, wearing out of insulations on old cables, improper control of voltage fluctuations, improper selection of motors, sudden high voltage whenever the power is resumed after a power failure are common reasons. Added to that, improper ventilation for motors results in fires.

The electrical lights and their wirings should be properly designed and enclosed. The problem is more where false ceiling is provided, as there shall be dust accumulation on the sheets and it is not easy to clean them daily. The fire on false ceiling spreads very fast all over the production hall.

In old factories, where cables are laid underground, and machines are getting modernized and speeds are increasing the load on the cables increases. There are number of cases of cable bursting resulting in major fires. All cables should be laid in such a way that they are accessible and there should be a record of the cable capacity and the load given to that cable. As the cables become old, the insulation cracks and wears out due to heat and dryness. Periodic checking of cable condition is very essential.

In case the complete mill is stopped due to power failure and power is resumed after sometimes there are chances of high voltage in the first machine started. The motor may burn or it may catch fire. Therefore after the resuming of power, phone to electric department and confirm that the power is stabilized. During monsoon season, because of high winds, trees may fall on the power line creating a short circuit and high voltage. The fires immediately after resuming power accounts for around 40% of the fire accidents due to electrical reasons in a textile mill.

1.9.5.3   Fires due to negligence of workers and staff

Almost all reasons for fire can come in the category of negligence of the staff and workers. If the maintenance is proper, if settings are proper, if safety regulations are followed strictly, if the housekeeping is good almost all fires can be avoided. There are number of bad habits which are leading to fire. Some common examples are keeping clothes, books and tiffin boxes on the control panels, keeping materials touching the electrical control panels, not maintaining the fire extinguishers, stacking materials near control panels, stacking inflammable materials near to wastes, smoking in banned areas, lighting lamps for making pooja in the production area and so on.

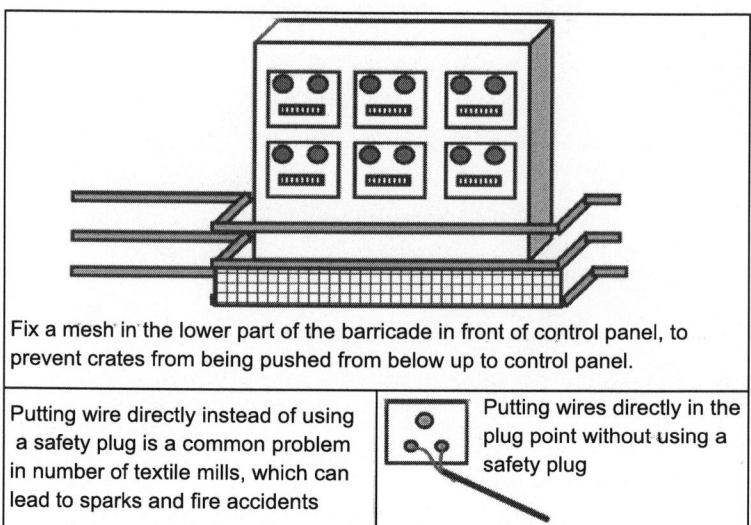

Fix a mesh in the lower part of the barricade in front of control panel, to prevent crates from being pushed from below up to control panel.

Putting wire directly instead of using a safety plug is a common problem in number of textile mills, which can lead to sparks and fire accidents	Putting wires directly in the plug point without using a safety plug

If the machines are run with higher speed than the capacity of the motors to take load, extra current flows in the winding. Then insulation breaks down resulting in short circuit.

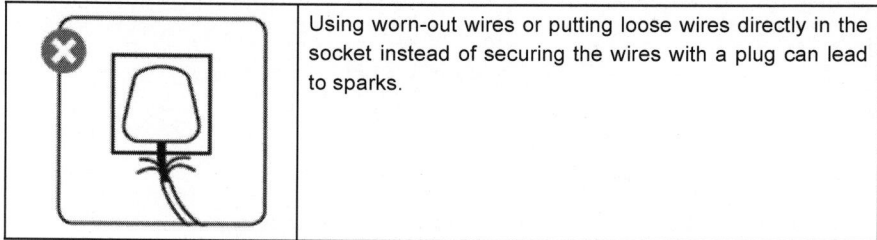

Using worn-out wires or putting loose wires directly in the socket instead of securing the wires with a plug can lead to sparks.

Allowing cotton dust to accumulate on the false ceilings, walls, corners etc., helps the fire to spread.

1.9.5.4   Fires due to nature's fury

There are some major fires due to nature's fury. A classic example is burning of over 2000 cotton bales which were kept in an open field due to lighting on 28th May 2005 at Gokak. The bales were kept in an open field, covered well with tarpaulins in lots of 100 bales each and sufficient distance was also maintained between lots. Suddenly in the night at around 11.35 PM lightning struck and there was a huge fire, which could not come down in spite of heavy rains. The staff and workers worked throughout the night in the heavy rains and shifted over 14000 bales to safe area.

There are instances of fire taking place due to static generation because of very low humidity in the atmosphere, and catching fire on the roof where dry dust has accumulated. Some such fires have taken place when the mill was closed due to power failure.

Earthquakes normally cause shaking of poles and wires resulting in short circuits. The lightening, trees falling on power lines, are normal especially in rural and hilly areas.

1.9.5.5   Precautions in case of fire

   i.   Every factory shall be provided with such means of escape in case of fire as may be prescribed like having minimum two fire exits in each shed with wide doors that lead to a safe place.
  ii.   In every factory, the doors affording exit from any room shall not be locked so that they cannot be easily and immediately opened from the inside while any person is within the room, and all such doors, unless they are of sliding type, shall be constructed to open outwards.
 iii.   Every door, window or other exit affording a means to escape in case of fire shall be distinctively marked in a language understood by the majority of the workers. Such marking should be in red letters of adequate size or by some other effective and clearly understood sign.
  iv.   An effective and clearly audible means of giving warning, in case of fire, to every person shall be provided in the factory.
   v.   A free passage–way giving access to each means of escape in case of fire shall be maintained for the use of all workers in the factory.
  vi.   Effective measures shall be taken to ensure that in every factory all workers are familiar with the means of escape in case of fire and have been adequately trained in the routine to be followed in such a case.
 vii.   It is far better to prevent the fire than it is to fight it after it has started. Therefore, the first steps in fire planning must include such things as machinery maintenance to assure that the machine will not start a fire by itself. Fire can start as a result of loose parts or poorly aligned bearings and shafts. If proper maintenance is practiced, most fires of a machine origin can be eliminated. This not only includes alignment of parts, but also proper lubrication and careful attention to electric motors.
viii.   It is suggested to have fire drills at least once in six months and all workers need to participate. A safe assembling area to be marked and head count to be done when workers assemble there. The time taken for the people to come out from the time of siren is to be noted down.

It is far better to prevent the fire than it is to fight it after it has started. Therefore, the first steps in fire planning must include such things as machinery maintenance to assure that the machine will not start a fire by itself. Fire can start as a result of loose parts or poorly aligned bearings and shafts. If proper maintenance is practiced, most fires of a machine origin can be eliminated. This not only includes alignment of parts, but also proper lubrication and careful attention to electric motors.

Although it is frequently difficult to have complete control of accumulations of waste, it is important that this "housekeeping" portion of fire prevention be observed. Periodic cleaning around machines is of extreme importance. Careful attention to broken wire bale straps occurring upon opening of the bales is extremely important. Broken pieces of wire bale straps are frequently introduced into opening machines, and will almost always start fires. This is the direct responsibility of the person charged with opening the bales prior to introducing them into the machines. Automatic fire protection systems on a Traveling or Circular Bale Opener will detect and suppress sparks and fire both inside the Bale Opener as well as on the cotton laydown.

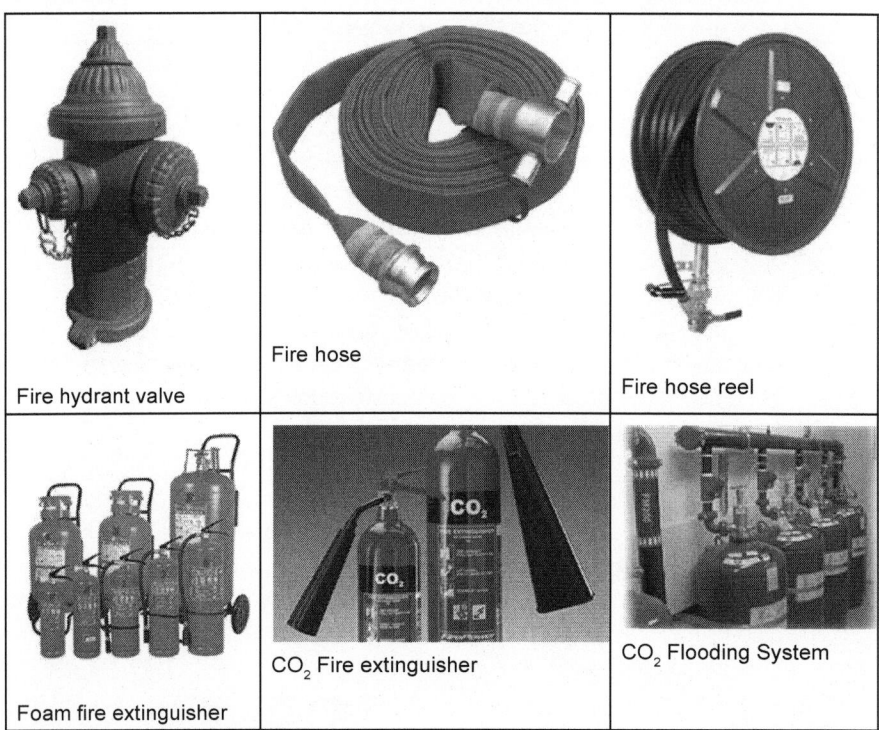

Fire hydrant valve

Fire hose

Fire hose reel

Foam fire extinguisher

CO_2 Fire extinguisher

CO_2 Flooding System

Removal of foreign ferrous material: All textile opener lines shall be equipped with magnetic separators, tramp iron separators, or other means for the removal of foreign ferrous material. This is very essential to prevent fire accidents in textile mills.

CO₂ flooding system: CO_2 flooding system shall be provided wherever cotton is being transported by pneumatic means.

Fire detection and diverting system: Wherever cotton is transported by pneumatic means, fire detectors and fire diversion system should be provided along with CO_2 flooding system to prevent fire from spreading.

Ban on smoking: Entire factory area of all textile mills, garment factories, cotton ginning mills, nonwoven factories shall be declared as "No Smoking Zone" as a preventive measure for possible fire accidents. No smoking sign board should be displayed at all prominent places in the factory.

Providing the fire extinguishers is the responsibility of top management, but maintaining them and periodically refilling is the responsibility of men on shop floor.

Poorly maintained fire extinguisher. The technical information is removed and no data of refilling.

Electrical control box is kept open.

Sprinkler system

Fire detection and fire diverting system in blow room

Water sprinklers supported by hydrant are to be provided at all places in spinning mills. The sprinklers open out in case of a fire and extinguish it.

| | Pump system for sprinkler | |
| Sprinkler controls and accessories | | Sprinkler system |

1.9.5.6   Online fire detection and extinguishing

Textile and nonwoven mills have experienced fires in the opening, blending, cleaning, carding, spinning, weaving, and filtration areas since the introduction of electrically operated machinery. In the past, these fires were controlled by alert mill employees who were able to bring manually operated fire extinguishers and fire hoses to the machine and fight the fire. Even if the fire was relatively large, it normally only damaged one or two machines. With the introduction of automated high speed production machinery and air filtration systems, the product moves from the Opening through carding processes without being touched by humans. The material is transported at speeds ranging from 10 meters per second to 25 meters per second. If a fire is started in one machine it can be transported to the next machine in less than one second. The fire can be spread through a complete Blow Room installation in less than one minute. This is normally too fast for the limited personnel in the area to react and stop the machinery so that they may fight the fire. Even if the machinery could be stopped quickly enough by an operator, the machines are normally completely enclosed in their housings, therefore making it impossible for a person to discharge an extinguisher or fire hose into the machinery. As a result of the high stock transport speeds, the enclosed machinery, and a reduced number of personnel in the area - when fire strikes an automated Blow Room installation it is possible to burn out several machines as well as the filters. With the increased cost and efficiency of the new automated machinery, most modern Blow Rooms have only two or three lines of this machinery. Therefore, when a line of machinery is burned out, the production of the Blow Room is reduced by one half or one third until the machinery can be repaired or replaced. This large drop in production underlines the need for serious fire protection planning by production-conscious mill managers.

Magnets and heavy material traps located in the stock transfer ducts are a way of reducing the problem of transporting fire-starting material in the stock ducts. A more effective way of eliminating metal is by using a high speed metal detector that will operate a pneumatic diverter. There are now metal detection and diversion systems available on the market that will detect ferrous and non-ferrous metal as small as 2mm and divert it into a collector box less than 2 meters away. These high-speed diverters occupy less than 4 meters of total duct space.

Even after all precautions have been taken, fires will continue to be a textile mill problem. These fires are caused by material "choke ups" in rotating machinery which will cause friction induced fires. It is almost impossible to prevent every fire-starting piece of foreign material from entering the process (such as stones), and even good maintenance practice will not prevent every machine malfunction from occurring. When a fire does occur, it is essential to detect the fire as quickly as possible and to control it. In addition to controlling the fire, the machinery must be stopped in order to prevent the passage of the fire from one machine to another, and alarms must be sounded to alert the personnel in the area.

In the early days of automatic textile fire protection, traditional smoke detectors were used in the belief they would be able to sense the large quantities of smoke generated in a textile fire. However the maintenance on these detectors, resulting from dust and lint entering the detectors, made their use prohibitive. These smoke detectors were not designed to operate in a lint-filled atmosphere, and later extensive testing was conducted on various types of heat, ultra-violet, and infrared detectors.

It was found that certain types of infrared detectors could successfully "see" a cotton fire moving at high speeds in a stock transfer duct. Since approximately 1980, infrared detectors have been accepted as the best of the available fire detectors for the textile industry.

These detectors have been improved and refined until they are now capable of sensing infrared radiation given off by small burning pieces of textile fibre moving at high speeds. Infrared Spark Detectors available today can sense a 1mm ember travelling at a speed of up to 60 meters per second.

Normally these detectors are located on stock transfer ducts in such a way as to be able to sense a spark or burning fibres moving from one machine to another. Upon detection of a fire in this manner a signal is sent to the central control panel to activate a Spark Diverter or an automatic 24 volt solenoid valve Extinguisher to control the fire. At the same time, a signal is sent to shut down the production line and filtration system so the fire will not spread from one machine to another. An alarm is sounded al the same time to advise personnel of the fire situation.

1.9.5.7 Fire control

The fire control systems normally used in textile machinery applications are Dry Chemical, Carbon Dioxide and FE-227ea. All of these agents are effective on textile fires when used properly. Each has advantages and disadvantages, and the decision on which to use should be made with these advantages and disadvantages in mind. However, none of these extinguishing agents can be guaranteed to completely extinguish any textile fire when a large volume of material is involved. These extinguishing agents are designed to control and prevent spread of textile fires. In most cases they actually extinguish the fire, although the user must be aware that some fire may remain in the machine after discharge. The material in the machine must be removed to make certain that no fire remains in the machine before restarting. In addition to automatic fire extinguishers, high speed spark diverters can also be located after the bale opener and before the mixers and chute-fed card lines to prevent embers from entering a machine. Used alone or in tandem with automatic extinguishers, spark diverters offer affordable fire protection for the blow room.

Advantages and disadvantages of various suppression agents

Agent	Advantages	Disadvantages
Dry Chemical (ABC)	• Fast Control • Relatively low-cost installation • Low-cost refill	• Requires clean-up time • Slightly corrosive • Leaves film on hot metal
Dry Chemical (BC)	• Fast Control • Relatively low-cost installation • Low-cost refill • Very attractive for Filtration Systems	• Requires clean-up time

Cont...

Agent	Advantages	Disadvantages
Carbon Dioxide	• No clean-up • Low-cost refill	• Requires stopping airflow • Potentially dangerous to personnel if inside machine or filter
FM-200/HFC-227ea	• No clean-up at all • Very effective fire control • Relatively low-cost installation • Environmentally safe • Very attractive for Blow Room machinery	• Requires stopping airflow • More expensive to refill

1.10   Adhering to social regulations

The industry runs successfully when the people in and around are supporting the industry. People within industry are the employees. The industry becomes successful because of the whole hearted support given by employees by way of hard working, applying their knowledge and experience for overcoming the problems faced by the industry, developing innovative techniques beneficial for the industry, working in a team to lift the company up. The employees are therefore termed as the most valuable asset. Similarly the people outside the industry are supporting the industry by providing food and shelter to the employees at affordable prices, running schools and colleges for the benefit of the children of the employees, providing transportation, providing security to the people, providing the infrastructural supports like communication systems, roads, water and power and so on. In other words, knowingly or unknowingly the industry is being supported by the community and the employees. It is therefore obligatory for the industry to support them and take care of them. Industry should be accountable to the society. How an organization proactively reacts to the need of the society is a good indicator of work quality.

1.10.1   SA 8000

Although the Factory Act prescribes the norms for employing people relating to non-employment of child labour, not employment forced labour, maintaining health and safety standards, freedom for formation of trade unions and collective bargaining, taking disciplinary actions, hours of working, paying overtime etc., the same are not followed religiously mainly because of the corrupt officials and greedy industrialists. Therefore the customers in the overseas market are under pressure of not accepting the materials from such companies who are not following the basic standards and are not accountable to society. Hence they are asking for SA 8000 Accreditation.

The Social Accountability Standards were created in 1997, which was created by Council on Economic Priorities Accreditation Agency (CEPAA) and later converted as Social Accountability International. 25 experts from various fields developed and drafted voluntary standards for social responsibility, which was first introduced in 1997 and then revised in 2001. It covers widely accepted international labour laws, which are in force in number of countries. In fact all the clauses of SA 8000 are already a part of Factory Act 1948 and all factories in India are supposed to follow it. However, due to inefficiency in the government to force the standards and immature short sighted policies and attitudes of industry owners, the standards are not being respected and followed. The Factory Acts, in fact are helpful to both the workers as well as to the industry. Because of the short sightedness of the mill owners, the industries are not able to retain the employees because of which they are not able to consistently produce the quality required by the customers. Although ISO 9000 insists on following all the legal and regulatory requirements, the part of social accountability is not taken seriously by the auditors while auditing a company for the compliance. Therefore SA 8000 is getting importance, as it specifically addresses the norms relating to Social Accountability.

There are 9 key standards in SA 8000 which are as follows.

1.10.1.1   Nine keys of SA 8000

1. *Child labour:* Manufacturers will not hire any employee under the age of 14, or under the age interfering with compulsory schooling, or under the minimum age established by law, whichever is greater. If a child of above 14 years or less than 18 years is employed, the management should provide time and facility for the worker to attend school and continue his education.
2. *Forced labour:* Manufacturers will not use involuntary or forced labour -- indentured, bonded or otherwise. The employees are free to leave the company if they are not interested in continuing.
3. *Health & safety:* Manufacturers will provide a safe and healthy work environment. Where residential housing is provided for workers, manufacturers will provide safe and healthy housing.
4. *Freedom of association and collective bargain:* Manufacturers will recognize and respect the right of employees to exercise their lawful rights of free association and collective bargaining.
5. *Discrimination:* Manufacturers will employ, pay, promote, and terminate workers on the basis of their ability to do the job, rather

than on the basis of personal characteristics or beliefs, sex, mother tongue, caste, the political group, etc.

6. *Disciplinary activities:* Manufacturers will provide a work environment free of harassment, abuse or corporal punishment in any form. Disciplinary action can be taken when an employee is proved to wilfully violate the code of conduct resulting in loss of property or causing injury or pain to fellow employees or management personnel.

7. *Working hours:* Manufacturers will comply with hours worked each day, and days worked each week, shall not exceed the legal limitations of the countries in which sewn or textile product is produced. Manufacturers of sewn or textile product will provide at least one day off in every seven-day period, except as required to meet urgent business needs.

8. *Compensation:* Manufacturers will pay at least the minimum total compensation required by local law, including all mandated wages, allowances and benefits

9. *Management systems:* Manufacturers will comply with laws and regulations in all locations where they conduct business.

The standards cover widely accepted International Labour Rights and address the Factory Level Management System Requirement. Independent, expert verification of compliance is done before certifying a company for SA 8000. The certificate holders are responsible for reporting to public regarding the activities they are doing and the system implementation. The system is harnessing consumer and investor concern.

1.10.1.2   Benefits of SA 8000

Benefits of SA 8000 are not only for the employees but also for business and consumers. They are as follows:

Workers & Trade Unions

- Enhanced opportunities for collective bargain
- Tool to educate workers
- Work directly with labour rights issues
- Public assurance for humane working conditions

For Business

- Putting companies values into action
- Enhancing company and brand reputation
- Improving employee recruitment, relation and performance
- Better supply chain management and performance

Consumers and Investors
- Clear and Credible assurance for purchase decisions
- Identification of products made ethically and companies committed to ethical sourcing
- Broad coverage of product categories and production geography

If the concepts of SA 8000 are really employed (not to take a certificate by somehow convincing the auditor as being done by number of companies) the relation between employees and management will improve and it leads to improvement in work quality.

1.10.2   ISO 26000

Influenced by the response to SA 8000 the International Organisation for Standardisation (I.S.O) developed guidelines under the title ISO 26000 which was released on 1st Nov 2010. This standard offers guidance on socially responsible behaviour and possible actions; it does not contain requirements and, therefore, in contrast to ISO management system standards, is not certifiable. As a guidance document the ISO 26000 is an offer, voluntary in use, and encourages organizations to discuss their social responsibility issues and possible actions with relevant stakeholders. As service providers, certification bodies do not belong to an organization's stakeholders. ISO 26000 encourages to reconsider an organization's social responsibility or "socially responsible behaviour" and to identify/select from its recommendations those where the organization could/should engage in contributions to society. ISO 26000 encourages further to report on actions taken. The need for organizations in both public and private sectors to behave in a socially responsible way is becoming a generalized requirement of society. It is shared by the stakeholder groups that are participating in the WG SR (ISO Working Group on Social Responsibility) to develop ISO 26000: industry, government, labour, consumers, nongovernmental organizations and others, in addition to geographical and gender-based balance.

1.10.3   ISO 14000 – Environment Management Systems

By considering the need for protecting environment, not only inside the working area, but also around the company, ISO developed guidelines for Environment Management ISO 14000 in the year 1996, which was amended in 2004. These International Standards are intended to provide organizations with the elements of an effective environmental management system

(EMS) that can be integrated with other management requirements and help organizations achieve environmental and economic goals. These standards are not intended to be used as non-tariff trade barriers or to increase or change an organization's legal obligations.

The standards specify requirements for an environmental management system to enable an organization to develop and implement a policy and objectives which take into account legal requirements and information about significant environmental aspects. It is intended to apply to all types and size of organizations and to accommodate diverse geographical, cultural and social conditions. The success of the system depends on commitment from all levels and functions of the organization, and especially from top management. A system of this kind enables an organization to develop an environmental policy, establish objectives and processes to achieve the policy commitments, take action as needed to improve its performance and demonstrate the conformity of the system to the requirements of this standard. The overall aim of this standard is to support environmental protection and prevention of pollution in balance with socio-economic needs. ISO 14001:2004 contains only objectively auditable requirements.

Willingness to protect environment by both management and employees leads to good work quality and quality of work life.

1.11   Material handling

The term material handling covers movement and storage of everything in and around an establishment. Proper material handling offers opportunity for improving productivity, reducing materials wastage, minimising industrial accidents, reducing man-power, etc. Material handling is often termed as necessary evil, as it does not add value to the product, but increases the cost, and sometimes damages the materials while handling. More handling means more damages especially in textiles. The slivers, roving bobbins and cones get damaged because of improper material handling.

Studies have shown that in textiles, manual handling causes more than a quarter of the work-related injuries reported each year. Around 60% of these involve an injury to the back, and some result in permanent disablement. Many injuries arise from stresses and strains over a period of time rather than from a single event. Manual handling problems often stem from poor workplace or job design. Among the most common examples of risky activities are jobs involving heavy or awkward loads, difficulty in gripping, excessive use of force, repetition, twisting and other awkward postures. Hence improving work quality in material handling is very important.

A high degree of automation is being propagated in textile industry for material handling which is less an issue of saving labour costs and more an issue of consistent quality. Because transport and storage require no labour involvement, there is no chance for human error and damages to materials and accidents to employees are reduced.

Provisions of lifts, hoists, suitable trolleys, railings etc., are essential for safe material handling. Providing the necessary equipments and accessories is the responsibility of top management whereas maintaining them and using them properly is the responsibility of people on the shop floor. In majority of cases, it is seen that improper handling of the material handling equipments is the main reason for accidents and damages.

2

Self-development and work quality

Quality of work can be good provided that our raw materials are good, machines are good, materials are good, methods are good and the people working are good. For making people good, it is not only the efforts of management but also the efforts of individuals to develop self are necessary. One who is enthusiastic to learn and implement new things can be trained and made good. If the individual has no interest, any time spent for training him becomes waste, as he is reluctant to learn. Man makes efforts to develop when he feels it as necessary. Unless the need is felt, people are not motivated to develop self. They will be happy in the same level where they are. Hence in order to improve work quality, management has to struggle to motivate people to learn and come up in their lives.

2.1   Need for self-development

All of us are working for one or the other organizations may be managed by self or from someone else. We may be working in or for a textile mill, a garment factory, ginning industry, trading house, research association, machinery manufacturing, educational institute, training school, hospital, business, charitable trust, temple or mutt. The organizations have an objective and we are working for fulfilling that objective. We should remember that we had an objective for ourselves when we decided to join or form that organization. We might have spelled it out or it may be hidden. Unless we aim for achieving that objective, we cannot be happy with the work we are doing for the organization. We need to develop so that we achieve our objectives and progress in that direction, then only the organization can develop. No organization can develop unless they provide opportunity for their people to develop. People do not stick to the organization where there is no scope for growth. Providing an opportunity for people to grow improves the quality of work significantly.

2.1.1   Objective of working

People are working to achieve their personal objectives. Working in an organization or society is a means of achieving personal objectives. By working in an organization we earn money by which we will be able to fulfil our objectives of looking after our family members, giving education to our children, leaving peacefully in own house, etc. If we have to achieve objectives, we need to work for it. We need to concentrate on the works to be done to achieve objectives.

People prefer working in an organization where payments are regular and good. If that organization has to pay regularly a good salary, it should earn substantially. To make the organization efficient to earn substantially is one of the main responsibilities of the people working. Therefore making the organization profitable should be one of the main objectives of all the people working followed by the objectives of the personal life.

2.1.2   Developing self to achieve the objectives

How can I make my company profitable? I should have the competency and powers to do this. Therefore I should make myself competent to do my works competitively. I have to watch the works I am doing and learn by that. I should develop a habit of reading and referring to the theory part of my work and develop logical thinking of making the work better. Unless I become strong, I cannot make my company strong.

When we talk of developing self we should consider physical fitness, mental ability to resolve the issues, knowledge to sort the problems, and attitude to mix with people and take them together. Unless we concentrate on all the four, we cannot develop. One should be disciplined in his food and other habits so that he does not fall sick or become weak. Mental ability develops as experience is gained and studying a situation with concentration and listening to the suggestions and feedback helps. Some people call it as maturity. For knowledge, one has to make efforts for learning from all. We can learn by discussing with learned people, observing the situation carefully and applying logics, studying the books and articles relating to the activities and the science behind it, observing the benchmark situations and understanding the difference between our system and good systems, being open and respecting the thoughts and logics of others and analysing the same with our situation, accepting that we are not perfect and there is a lot to learn and making continuous efforts for learning. Attitude to mix up with people comes when we start developing the habit of recognising strength and good parts of others, instead of seeing only their weakness. Becoming a member of professional association and taking active part can help people to grow. There are number of such associations: to name a few "The Textile Association of India", "Institute of Engineers", "Indian Society for Quality", "Asian Network for Quality", "All India Association of Textile Chemists", "The Textile Club" and so on.

2.2   Ambition and motivation

All on earth have ambitions, but some express and others do not. Some work for achieving that whereas others enjoy only in daydreaming. If one feels that he is capable of achieving that ambition, he shall be motivated and uses all opportunities to achieve that. One who achieves his ambition shall be one of the happiest on earth.

2.2.1   Needs and ambitions

Needs are necessary to be fulfilled, whereas ambition if fulfilled makes man delighted and happy. As everyone knows, the basic needs are food, clothes and shelter. If a man is struggling to meet his basic needs, he cannot concentrate on fulfilling the ambitions. If the management does not stand up and ensure that basic needs of the employees and their family are met, they cannot expect those employees to respond to the ambitions of management of running the company successfully and making profit. Hence work quality shall not be achieved if basic requirements are not met.

If we take the example of a textile mill, producing the quality as needed by the customer, delivering the materials as per customer's requirement and producing the material at affordable cost are basic needs. Being No. 1 in the industry, winning various awards, earning a name as a philanthropist, getting membership of various professional clubs are ambitions. The industry should first concentrate on basic needs, for which they need to device systems and adhere to it. Cleaning, housekeeping, taking safety precautions, preventive maintenance, training people to do their assigned works properly, responding to customer's complaint and feedbacks, analysing the performance, taking proper corrective and preventive actions, understanding the customer needs and requirements and aligning all the activities, employing appropriate technology for the products being manufactured are essential steps for surviving of an industry. Industry's ambitions can be fulfilled only when the needs are fulfilled, i.e., industry surviving by providing quality products and services in time at competitive prices.

2.2.2   Need for fulfilling ambitions

Is it really necessary to fulfil the ambitions? Is it not possible to survive without fulfilling the ambitions? If surviving is the only goal, somehow one can survive, whereas when ambition is the goal, one has to plan and organise the activities and work for continual improvement. The ambitions develop interest among people to achieve something for which they strive. Although one cannot fulfil the ambitions, the efforts made make him much better than what he was. For example, a person having ambition of winning an Olympic medal shall practice daily and move much upwards than the level in which he was. Similarly the ambition of being No. 1 in the industry, makes the people in the organization to concentrate their efforts towards a common goal and the organization moves forward, and shall become strong and competitive. The target of only surviving might make the company sick as the market favours leaders and winners, and they overtake the companies only targeting for survival. Hence having an ambition of growth and working for it helps in improving the work quality.

2.2.3   Motivating self to fulfil ambitions

Although all have their own ambitions, they are not motivated enough to move in the direction to fulfil their ambitions. The failures seen in life are the major demotivating factor. We need to analyse the reasons for failures and make plans to prevent those factors from coming up again.

One should understand that the world does not remain same, and there are always many developments. The factor responsible for earlier failure might not exist now. There may be number of factors that may be helpful to you in achieving your ambition but you may not be aware of the same. You also are not going to remain same; your knowledge, experience and maturity improve, and you will certainly do not allow those factors to creep up and cause a failure. Count on the successes you have and you will realise the strength you have and the potential to achieve the ambition. Even though you cannot achieve the ambition fully today, but by moving in that direction steadily, you are going to achieve it.

How can you motivate others in your team to achieve the ambitions of the company? They have their own perceptions and fears. They highlight the failures they faced in attempting to achieve this ambition. They have come to a conclusion that improvement is not possible in this organization. You need to take small projects one at a time and involve people in making that successful and give credit to the people worked and highlight their achievement. This builds confidence and enthusiasm. Once the people are confident of achieving, filling enthusiasm is not a difficult work. So, identify the strengths of your people and allocate the works which they can do successfully and encourage them with due recognition after each success. These develop enthusiasm, and at that time gradually attack the weaknesses one by one.

2.3   Learning and development

We may motivate people to do a job, but unless they have the required attitude, knowledge, and competency they will fail. One should have the correct understanding of the system, the knowledge of the science behind that operation, the awareness of the supporting and conflicting factors and the situation as on date. In the schools and colleges the basic principle of operations or the science behind an activity shall be taught, whereas others are to be learnt while working.

2.3.1   Analysing the success and failures

The best teacher is the experience. We normally do not forget whatever we learn from experience, but may forget what we studied in the school or college which are not being used in our regular works.

While working we get both success and failures. In number of cases we attribute it to luck and fate, but we never question why the luck or fate should favour at a particular situation and not at all the times. It is necessary to analyse why we failed or got success. In our success, sometimes we may recognise a luck factor, like opponent could not come in time due to some disturbances in the road, heavy floods blocked materials coming from outside and our materials could be sold in the market, a large organization selects one of our shirt design for their employees uniform and we get a big order in that design, and so on. Similarly, sometimes we may identify a fate factor in our failures like the raw materials did not reach in time due to strike by truck drivers, the common effluent treatment plant of the industrial estate failed, because of which we had to shut down our process plant, and so on.

Whatever may be the case, we need to analyse and record all the happenings, whether we accept it as the reason or not. Go on recording all the events, which shall help you, like a CCTV camera, recording all the events, and you can analyse later.

2.3.2   Learning from success and failure

When we analyse our successes and failures, we can learn from both. My decision of increasing flat strips and comber noil to reduce imperfections was a failure as I did not have the basic culture of keeping the machines and surroundings clean, and maintaining the machines with proper cleaning and lubrication was lacking. In such a situation, my decision of increasing in flat strips and comber noils added to the cost, but results were not achieved. Another my decision of increasing the speed to get higher production was also a failure. It resulted in higher breakages, higher wastes, higher imperfections and more stoppages for attending breakages. I might have done a mistake of taking wrong decision, but when I analysed, I could understand my fault. I will not do the same mistake again. It is learning from failure.

Let us take another example. In a particular month, I achieved the production more than the target given to me. Everyone congratulated me, but I did not know the reason why I got more production. Whether it was due to the combination of raw materials or the combination of product mix which led to well-balanced working, or the weather conditions that was conducive or the availability of skilled workmen as there were no festivals or leaves or some other factor which we were not able to identify, which is not clear to me. Unless we are able to identify the real reason for a success, we cannot repeat it. Then it is our luck rather than our competency which brought the results. In number of cases it is seen that rather than the contribution of a single factor, a combination of number of factors would be the real reason for success. We can repeat the result provided we can bring all the combinations together.

Analysis of failures indicates what we should avoid, whereas analysis of success indicates what we should adapt. We should learn from both.

2.3.3   Sharing the learning

By only learning we cannot succeed, as what I learnt is not learnt by my team members. So I need to educate them by sharing my learning. By sharing the learning there is no fear of me loosing anything, as my learning improves by answering number of questions and doubts raised. My team members become more friendly with me as I share my learning or experience, which they could learn something. Once the team becomes strong and cohesive, we get more strength and will be able to achieve our targets and fulfil our ambitions. Our work quality is bound to improve.

2.3.4   Take challenges

Monotony in work makes the work boring. Human beings like change; they do not want to wear the same type of clothes all the time, do not want to eat the same food all the time, do not want to hear to same music, do not want to see the same movie. Similarly people do not want to do the same work all the time. Man enjoys taking challenges; however, he may hesitate when the management asks him to take higher job or a challenging job as he is accountable. Once that challenge is faced and man comes out successfully he shall be very happy and enjoys the work.

In number of cases, the management may not ask us to take challenging job as they are more interested in routine jobs and would like us the complete the routine jobs in time as per their plan. People feel bored as doing the same work repeatedly is not interesting; but an enthusiastic person shall identify the challenges in the routine job itself and overcomes them by modifying his method of working or by taking precautionary measures. This leads to improvements. So one need not keep quite stating that job is monotonous; think and work, you can see challenges in that. Overcome the challenges and grow.

Take simple examples of reducing the time for doffing in a ring frame, reducing the time for loading or unloading of materials, maintaining the clean floor in a spinning shed, allocating winding drums as per the packing requirement, keeping the materials in their locations in stores, motivating fellow workers to do their jobs in time, counselling the absenting workers and making them attending regularly and so on. All jobs have challenges in them. See where the expected results are not coming, see where you are falling behind compared to your competitors, see where you can challenge the status quo and work on that. Your job shall become interesting and quality of your work improves.

2.4   Leadership and maintaining discipline

Each one has lots of desires and ambitions and makes efforts to achieve them. One feels as an achiever if he can meet those. People take leadership for achieving their objectives. If you want to win, you should take leadership for achieving your targets. You cannot leave this responsibility to someone. We cannot get the results wanted by us under the leadership of others. When I work under the leadership of others, I get the results that others wanted to achieve. If others work as per my wishes, then I shall get the result wanted by me. Hence I should become leader for my activities. My work quality will improve when I take leadership.

There are a number of instances that no other person could do our job, but we are not starting the job at all. We are ready with excuses for not doing that work, whereas we not even make a simple attempt to think on how to do. When someone compels, we do the work. We respect the one who compelled us to do our work, and recognize him as "Guru" or "Leader". There are numerous examples for this, like getting up from the bed in time, keeping our work area clean, keeping our belongings such as books files records in an organised manner, attending to meetings or training related to our works in time, documenting our activities for self-assessment, following diet and doing exercises for maintaining self-health, planning owns activities, having self- discipline, etc. We need to wake up and take leadership to initiate our activities. One should remember that no work can be completed unless it is started. So we should start the work. We should take leadership for doing the work.

Maintaining self-discipline is very important for anyone, especially for a leader as people follow the deeds of a leader and not the words. Result in the work can be achieved if all the activities are done in the way it is to be done. It is called as discipline. If any step is missed, there are chances of failure.

When Lord Sri Rama became king, his guru Sri Vasishta told "यथाराजतथाप्रथा" (*Yatha Raja Tatha Praja*) meaning that the people shall follow the footsteps of the King. If King is disciplined and true to his words, the citizens also shall be disciplined and true to their words. Sri Rama took this advice seriously, and was very cautious in all his deeds, and became a model, and was called as "मर्यादापुरुषोत्तम" (*"Maryada Purushothama"*) i.e. highly disciplined Super Human. This is true for all the time. In any society, organization or institution, the people follow the steps of their seniors, just like small children trying to imitate and follow the steps of their parents or elder brothers and sisters, and their teacher.

2.4.1   Defining self-norms in line with the work and society

Our elders have defined norms for various activities and we are following them with great respect in the name of religion or culture and are not ready to deviate. Our elders could develop norms for the activities which were familiar to them. They studied them deeply before arriving at a conclusion. They followed it strictly which were followed by disciples. They were so strict in following the discipline, it all developed as religion and culture. For the systems and works developed later, who should prepare the norms? Who is interested in developing norms and who gets benefits from following norms?

The one getting benefits or one getting affected shall normally be interested in the norms. Unfortunately, in number of cases, people are expecting someone else to develop norms whereas that someone is a third person and not having any interest in the subject. The research associations, standard bodies and consultants are doing the work of developing norms in textile industry, whereas the people working in the industry do not have any idea as to how the norms were developed. They are forced to achieve results as per norms and are failing.

2.4.2   Developing methods to achieve

Only by writing norms or by-hearting the norms, we cannot achieve. We need to develop methods. A work may be done by using different methods, but we need to work out which shall be suitable for our present situation. We can get good quality yarn by reducing the speed at carding; but whether I will be able to fulfil the production requirement. I can get better production by employing more doffer boys to help taking doffs faster and to help the siders in piecing; but whether I will be able to maintain cost of production. Sometimes by employing more people for a job than required, I shall be making people less efficient and leading to more confusion. A large inter bobbin helps in reducing material handling, improves efficiency in speed frames as well as in ring frames, but this theory might not hold good if I am spinning superfine counts. Installing latest high speed looms may give higher production, but not when I am taking small lengths of sized beams. By not permitting a worker coming late by 10 minutes for work might help in improving discipline, but I will be losing production for the day due to short of worker. Switching some of the lights might reduce the power bill, but may allow the defects to go in fabric.

We need to study the method and ensure it as the appropriate method for getting the expected results. Develop the procedure which can be implemented by all in the section. Identify the areas to be controlled to get the results and have checks. Educate the people on the method you want them to work. Demonstrate the method and convince that you can get the required result. Involve the people actually working for developing methods so that they can feel the ownership.

Developing methods are not only restricted to the core work you are doing, but is applicable for other related activities also. You need to device the method of explaining to your juniors and colleagues regarding the importance of work and the way to do so that they are convinced. You need to develop methods of reporting to your seniors in a way that they could see the good points in the work done and accept it. You need to develop the method of recording so that you get the relevant information when you need them for analysis. You need to develop methods for educating and convincing the customers to recognize the good points of your work and adapt it as their norms for procurements.

2.4.3   Following religiously

Following the system religiously is more important rather than taking short cuts to get the results. By driving in a wrong side, I might reach faster, but chances of meeting accidents are there. If I do not meet accident, it is luck. Spending 10 minutes to clean the machine and surroundings is much better comparing to running the machine without cleaning and ending up with poor quality and breakdowns. Discussing with the outgoing worker for 5 minutes before starting the work is much better rather than ending up with a problem. Religiously following colour codification is much better rather than mixing different colour bobbins to get more production. Religiously checking the incoming materials at stores is much better compared to using them and ending up with a problem. Religiously checking each spindle for quality is much beneficial rather than allowing a bad spindle to bring market complaint. Religiously taking round by a senior manager is more useful to the company rather than advising the people on the theory of getting good quality. Religiously discussing the problems with juniors is much better rather than assuming everything as good. Religiously checking wastes is more beneficial than allowing good materials going in waste. If you are religious in following systems, your assistants also shall follow and you will get the result. So follow your preaching religiously and ensure that your work quality is achieved.

2.4.4   Developing leadership to achieve

You need to develop and demonstrate your leadership in achieving the objectives. You should be proactive in doing works and not wait for anyone to remind you. In number of mills, the senior officials have developed a habit of not responding unless a reminder is given. They give the logic that "If it is really important, I should be reminded as I will be busy in so many works". However, the real achievers will not wait for a reminder, but will be doing the work as per schedule.

We normally hear that "Here unless it is followed, work will not be done. We have to be after them. The management should provide more supervisors to do the following up of jobs, and a senior manager to follow up with the supervisor. The MD has to follow up with senior Managers...... and so on." Sometimes we get blamed by the seniors when we remind them for not responding to letters stating that "Do you mean to say that I should see all letters and emails. I am too busy. If anything is there, come and tell me personally". When a meeting is called and circular is sent well in advance, the members shall be expecting a phone call from the convener half an hour before meeting. These are all bad habits of senior people, and their subordinates simply follow the elders, and never are proactive in doing their jobs by themselves. Such companies cannot survive for long. Such companies are surviving only by concessions and benefits given by the governments, not following the regulations and not paying the taxes by keeping top government officials in good terms. There is no work quality in such organizations.

If you have to progress and achieve the results you wanted, you need to be leader for yourself, and lead your life as per your plans, and maintain the discipline essential for achieving the result.

2.5   Quality of working life and its assessment

"Quality of working life" is a term that had been used to describe the broader job-related experience an individual has. Earlier researchers had concentrated on the term "job satisfaction" and suggested that if a person is satisfied with his job, he can be happy and his quality of work shall be better. However in recent years, although a man loves his job, the stress in the job is considered as more dangerous. A highly enthusiastic person who likes his job takes more responsibility than what he can perform with ease and feels happy while achieving it, whereas in the process he undergoes lot of stress that can spoil his work life. The precise nature of the relationship between the concepts of stress and subjective well-being has still been little explored. Stress at work is often considered in isolation, wherein it is assessed on the basis that attention to an individual's stress management skills or the sources of stress will prove to provide a good enough basis for effective intervention. Alternatively, job satisfaction may be assessed, so that action can be taken which will enhance an individual's performance.

Quality of work life is specifically related to the level of happiness a person derives from his career. Each person has different needs when it comes to their careers; the quality level of their work life is determined by whether those needs are being met. While some people might be content with a simple minimum wage job as long as it helps pay the bills, others would find such a job to be too tedious or involve too much physical labour and would find such a position to be highly unsatisfactory. Thus, requirements for having a high "quality of work life" vary from person to person. Regardless of their standards, those with a high quality of work life generally make enough to live comfortably, find their work to be interesting or engaging and achieve a level of personal satisfaction or fulfillment from the jobs that they do. In other words, employees who are generally happy with their work are said to have a high quality of work life, and those who are unhappy or unfulfilled by their work are said to have a low quality of work life.

Quality of working life has been differentiated from the broader concept of quality of life. Elizur and Shye, in 1990, had concluded that quality of work performance is affected by quality of life as well as quality of working life. However, it is argued that the specific attention to work-related aspects of quality of life is valid. A clearer understanding of the inter-relationship of the various facets of quality of working life offers the opportunity for improved analysis of cause and effect in the workplace. This consideration of quality of working life as the greater context for various factors in the workplace, such as job satisfaction and stress, may offer opportunity for more cost-effective interventions in the workplace.

Hackman and Oldham in 1976 considered psychological growth as relevant while considering quality of work life. Taylor (1979) identified the essential components of quality of working life as basic extrinsic job factors of wages, hours and working conditions, and the intrinsic job notions of the nature of the work itself. He suggested that a number of other aspects could be added, including individual power, employee participation in the management, fairness and equity, social support, use of one's present skills, self-development, a meaningful future at work, social relevance of the work or product and effect on extra work activities.

Warr and colleagues (1979), in an investigation of quality of working life, considered a range of apparently relevant factors, including work involvement, intrinsic job motivation, higher order need strength, perceived intrinsic job characteristics, job satisfaction, life satisfaction, happiness, and self-rated anxiety. They discussed a range of correlations derived from their work, such as those between work involvement and job satisfaction, intrinsic job motivation and job satisfaction, and perceived intrinsic job characteristics and job satisfaction. In particular, they found evidence for a moderate association between total job satisfaction and total life satisfaction and happiness, with a less strong, but significant association with self-rated anxiety.

Mirvis and Lawler suggested that quality of working life was associated with satisfaction with wages, hours and working conditions, describing the "basic elements of a good quality of work life" as safe work environment, equitable wages, equal employment opportunities and opportunities for advancement. Baba and Jamal listed typical indicators of quality of working life, including job satisfaction, job involvement, work role ambiguity, work role conflict, work role overload, job stress, organisational commitment and turn-over intentions.

An individual's experience of satisfaction or dissatisfaction can be substantially rooted in their perception, rather than simply reflecting their "real world". Further, an individual's perception can be affected by relative comparison. There are few recognised measures of quality of working life and jobs. Regular assessment of quality of working life can potentially provide organisations with important information about the welfare of their employees, such as job satisfaction, general well-being, work-related stress and the home-work interface. Some of the factors used to measure quality of working life pick up on things that don't actually make people feel good, but which seem to make people feel bad about work if those things are absent.

3

Customer orientation and work quality

Quality is defined as fitness for the purpose of achieving customer satisfaction, while adhering to legal and regulatory requirements and ethics. In a business we have the prime objective of earning so that we can run the business and survive. For this purpose, we need to understand the customer needs and all our activities should be oriented to meet that objective of fulfilling customer's needs. My work quality should be in line with achieving customer satisfaction.

3.1 Concept of customer

Who is our customer is a question generally not answered properly. For an organization, one who is paying money and purchasing goods and services is normally referred as "customer", whereas some other say "when you are delivering something, the one who is receiving is your customer – there is no need that he should pay you money, he may be paying a lot to you to make you strong". This statement is difficult to digest. If those are within our organization, we refer them as internal customers, but there are people outside the organization and we are delivering them some of our products and services in some form.

3.1.1 Customer as a stakeholder

The stakeholders are those who have interest in us; they want us to grow because of which they also can grow. When we talk of a business organization we need shareholders and financial institutes to fund our organization to start. We need the government to provide all facilities like infrastructure, tax holidays, licenses, etc., and the employees for working for us to produce and provide the services to our customer. The suppliers are needed to supply us the required materials and services to run our business, and the community to support us all the time in all our activities. We need customers to purchase our products and services and pay for the same.

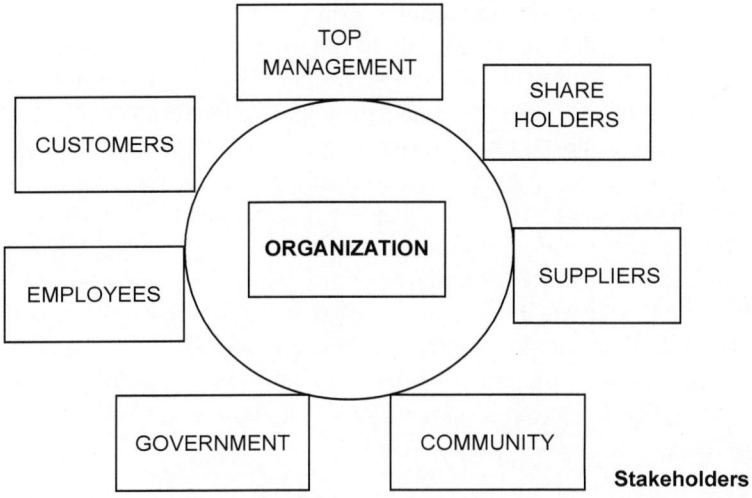

Stakeholders

The customers are considered as most important stakeholder as they purchase our products and services and pay for the same. If they do not purchase, whatever is invested and produced shall be a waste. All quality management systems are centred on understanding and fulfilling customer requirements. They insist that all activities should be customer focused and everyone in the organization should work to fulfil the customer needs.

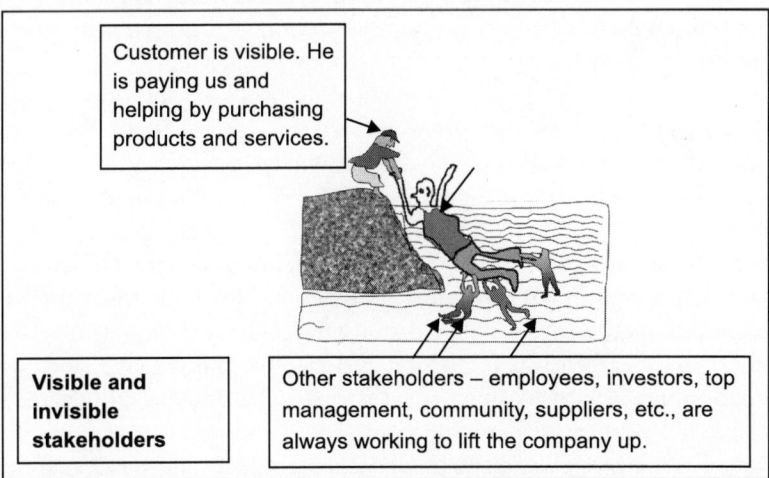

The customers pay money after getting the product and services; whereas the shareholders invest money in our business without getting anything from us in advance. The top management plans and provides resources to achieve their dream targets whereas the employees make it a reality. The suppliers supply the required quality materials in time at affordable prices so that we can produce the materials and supply to our customers. The government and community are providing various supports and infrastructure for us to succeed. The customers are visible as the one lifting the organization out from sinking by paying money and purchasing the products and services, whereas others push the organization up by going down in the deep water. The customer is standing in a safe place, whereas others are in deep waters. They will come up only after the organization is pulled up by the help of customer. If customer feels that he cannot lift this organization up, he can just leave it and go out, whereas others, who are already in deep waters, cannot leave the organization there. They shall also sink along with the organization.

In order to ensure that the organization does not sink, we have to improve our work quality, by which products and services are improved and customer stays with us.

3.1.2   Customer as a provider of opportunity

Mahatma Gandhi, much before he could be recognised as "Mahatma" in 1890, while addressing a gathering of businessmen at South Africa said that "A customer is the most important visitor on our premises. He is not dependent on us. We are dependent on him. He is not an interruption of our work. He is the purpose of it. He is not an outsider of our business. He is part of it. We are not doing him a favour by serving him. He is doing us a favour by giving us the opportunity to do so."

Customer is not just paying money; he is providing an opportunity for us to grow. He makes us strong and experts by forcing us to plan, design, develop and practice by which we gain knowledge and practice and become strong and confident. Take an example of a fabric designer. Unless there is a customer for him, his talents and creativity will not get exposed. Customer may want one design, but the designer makes ten out of which one is approved. Designer gets a practice of designing new things. When customer insists for the benchmark quality, we strive and study the latest techniques, observe the benchmark partners and develop our system. The knowledge and system developed remains with us making us strong, whereas the money received might have been spent for something else.

Customer has money and can purchase the products and services from any supplier; then why he should come to us? If he feels that we are reasonable in our dealings, provide the quality and services in time, he shall come to us. He has innumerable number of suppliers, and can purchase from any one, but for us he is the only one to whom we should serve and make him happy so that we get repeat orders. We have to respect and serve the customer who has come to us rather than daydreaming another big customer, who has not come to our door. The customer standing in front of us and asking for materials and services is worth much more than any customers on earth, as others have not come to us, and we have no guarantee that they will come to us. Our quality of work in dealing with customer attracts the customers more than the quality of product.

3.1.3   Customer as a critic for improvements

The customers when they are not happy but have a confidence on the supplier shall make complaints. If the customer is not having confidence on the supplier, they shall change over to another, as complaining and fighting for compensations is costlier compared to changing over to a new supplier. It is observed that only 10–15% of the customers make complaint whereas the remaining prefers to change the supplier without making any noise. When customers are not complaining, it does not mean that they are happy, as we have supplied same quality to all, and that is our quality.

The customers complain or express their unhappiness not only on the quality of the products but also on the services that include the way in which the customers are treated, their queries and calls are responded, information provided, the packing and forwarding, etc.

No customer is interested in complaining, but is interested in running his business, and he wants a smooth working. Customer has to spend for making a complaint. He has to collect evidences to prove that the materials supplied or the service provided by us is of substandard nature. He has to correspond with us, has to wait till our representative visits him and analyses the problem, has to provide us the facilities to study the problem at his place. Normally customer prefers to purchase from a reliable supplier who can assure the quality in the first attempt itself and not the one who analyses the problem after a failure and gives oral assurances that the quality shall be taken care in future. If a customer is taking pain in sending a complaint to us, it means that he want us to improve. He provides the information where we are weak so that we can strengthen ourselves.

We should never be upset with customer who makes a complaint, but take it as a positive move to improve ourselves and go back to him with improved results. We should request customers to indicate our weaknesses so that we can improve and not be happy whenever there is no complaint. We should rather be worried if we do not get any complaint from a customer, as he may be searching some other supplier and ignoring us. Only 10–15% customers complain, who have some faith in us and our approach. So we have to take customer as a critic for improving our quality and systems.

The companies appoint consultants to study their systems and products and suggest methods for improving it. The consultants charge for that. The customer does not charge for the feedback given, and also shall be happy to work with you provided you are proactive and sincere in solving the problem, which is affecting both you and the customer. A consultant might exaggerate the things so that he can earn more, but the customer shall point out what exactly is missing or harming him. A consultant need not be very keen on the result, but on his continuity as a consultant, whereas the customer is keen on results. No customer can make profit by putting claims. He shall be losing a good supplier and finally gets only poor quality. Our work quality should address understanding the customer clearly and aligning with the customer needs.

3.2   Internal customers

The one who purchases our products and services is called as customer. Unless the customers purchase, our products and services has no value. Customers provide funds to run our business by purchasing. Customers show the direction in which we need to move.

We are providing products and services to our next man in the organization, who is termed as an internal customer. Internal customers need good quality and services so that they can produce good quality products to give to external customers. We should ensure that our internal customers are satisfied with the quality, delivery and services. External customer cannot be satisfied unless internal customer is satisfied. Our work quality should address the needs of our internal customers.

3.2.1   Identify your customer

It is necessary that you need to identify your customers, who takes your products and services. For a person in production line, the next person is the customer. If draw frame person is satisfied with the quality and supplying in time of card cans, he shall be satisfied.

If you are a supervisor, both the people above you and working under you are your customers. You report to your boss. Your boss is receiving your report. He should be able to understand what you want to explain. If your language is good, you shall be able to write clear sentences so that boss can understand. If you make mistakes in spelling and because of that the meaning is getting changed, there shall be lot of confusion. He may take a wrong decision because of your mistakes in report. You are giving instructions to your assistant, who is less educated than you. He may not understand. When he cannot understand your instructions, he cannot do the correct work.

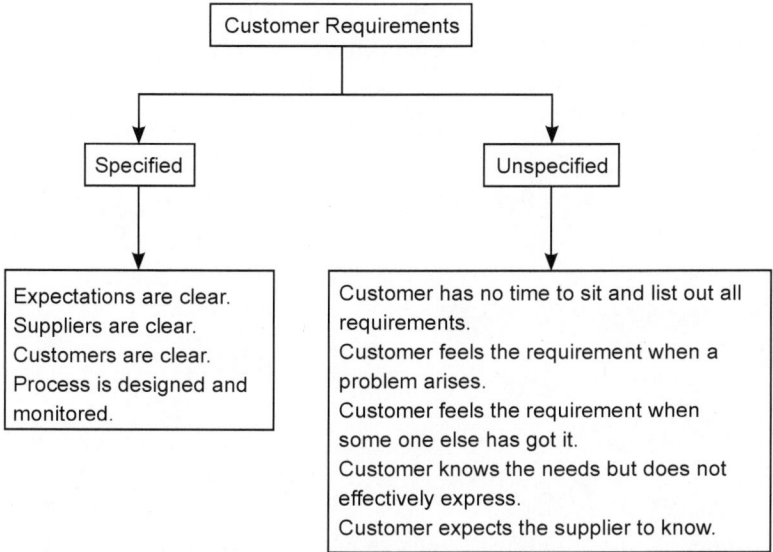

When you are in a meeting discussing some point, all others listening to you are your customers. When you are sending a circular, all those read it are your customers. When you write instructions in a log book, all those who read it are your customers. It means, in your company all are your customers. So develop the habit of providing what is required by them. Your top management wants productivity with least expenses; give it. Your boss wants you to salute him in front of others; salute him; you are not losing anything. Your next process wants quality with timely delivery; work of it. Your assistants want your support in solving departmental problems; give them support. Your colleagues want your cooperation; give them.

3.3   Understanding the customer needs

There are number of expectations from the customer depending on the product and services being procured and the purpose of procuring. One needs to discuss with the customer, understand the purpose for which the material or service is being purchased, the way in which it is used, the culture of the people concerned, their likings and disliking, the social obligations, the legal and statutory requirements and the price customer can afford. The supervisors should discuss with the marketing personnel and get the requirements that are specific to the order. While doing this, the intended use and the quality needs considering the objectives of the product are to be understood. In number of cases, the specifications given by the customer shall not be complete, and in such cases, the technicians have to complete the specifications as per similar products and get the approval from the customer through the marketing section before starting the production. It is also customary to give a prototype sample to the customer and get feedback and finalize the specifications. In garment industries, the samples are produced at different levels and got approved by the customer before starting bulk production.

The general expectations from the customers are as follows.

- *Desired products* – The customer pays and buys the product required by him; and hence, we need to produce the materials as per his requirements. We might have produced something fantastic, but he cannot purchase unless it is needed.
- *Timely delivery* – The customer needs materials for using at his end at the time when it is needed and not when not needed. It is a waste if not available in time. Non-receipt of material in time can become a threat to the running of business itself; customer might have to lay off the people for no work.
- *No increase in prices* – The customer shall have planned his activities considering certain costs for the materials being procured. If the prices are increased in between, all his calculations shall fail, and he shall have to face losses. Hence no increase in price is accepted.
- *Prompt and quick service* – The customer purchases materials from us to use them. While running the materials if he finds any problem, his activity shall be affected, and hence he wants our help. We need to provide timely service.
- *Smooth working at his place* – The materials are purchased to work smoothly at the customer's place, and not to create problems. Hence, while deciding the process parameters and product specifications, we need to understand the purpose for which the materials are being taken and design the product accordingly.
- *Compensate for the losses due to quality* – Customer has purchased materials to run his business. Why he should suffer because of the quality problems in our supplies? He demands for compensation much higher than the sale value of the supplies. Hence, we should be clear on the objectionable faults or errors that can happen and design systems to prevent.

3.3.1   Stated needs

The customer needs are innumerable. Some are expressed whereas others are taken as granted. The stated needs are the one expressed in writing in the purchase order, which include the product descriptions, product specifications, drawings and representative samples, quantity to be supplied, the delivery schedule, the rate per unit and the applicable taxes, the transport and insurance, type of packing, the destination, etc.

3.3.2   Implied needs

The customer normally tries to provide as much information as possible, which he feels as important, but there are number of needs, those are implied and taken for granted as understood. If we take an example of yarn, the customer specifies the count, twist, minimum strength, maximum imperfections, weight per package, number of packages per carton, etc. It is understood that the cartons are tightly packed, the information written on cartons are neat and legible, the cartons are not dirty, the winding is good, the cone quality is good, the cones are covered with polythene covers, suitable separators are put to avoid cone damage, the invoice and challans are legible, the invoices are signed by authorized persons, the address and names written on invoice are clear, the quantity written in the invoice and challan are same, the labels put on the cone are of the same yarn, all yarns in the lot are made from same mixing lot, there is no shade variation between cones or within cones, and so on. Everything cannot be specified, but the supplier should understand and react.

Customer feels the requirement when he faces a problem. Normally customers allot a product code and all information about the product is given in as much detail as possible. If we analyse, the stated needs are very less compared to implied needs. It is like a tip of an iceberg. That is the reason, when we analyse the complaints, 94% of customer dissatisfaction are not due to product quality, but due to various other reasons. Hence good quality of work includes understanding the unstated needs and fulfilling them.

3.3.3   Quality, delivery and timely service

As quality of product is important, the quality of delivery is also important. Quality of delivery includes supplying required quantity at required time. If customer has given break up of delivery with different quantities, we need to adhere to that. The packing, in other words, presentation should be good and precautions should be taken that materials are not disturbed during transit.

Early delivery is not suggested as customer shall not be able to keep the material in his warehouse, and may not be able to make arrangement for finance to give you. Late delivery is creating huge loss to the customer, as his men and machines become idle for want of materials. In case of exports, delay in a day or two may result in missing of a vessel, and we may have to wait for weeks together to get a vessel to that destination, or send the material by air, spending huge amount. Hence a good quality of work includes doing the work in time.

3.3.4   Affordable costs

Affordable cost is one of the prime requirements of a customer. If our products are not affordable by the customer, he will not purchase, whatever may be the quality benefits. A combed soft yarn can make a better towel than a carded yarn, but customers prefer carded yarn as it is cheap. Similarly a combed yarn can make a better T-shirt, but majority of customer prefer carded yarn as it is cheap. If same quality of yarn is offered by two spinning mills, and if the rate difference is there of 1 Re. 1 the customer prefers that yarn with low price. Unless we have built a brand image and customers are confident because of the brand, one can quote slightly higher price and not a very high price. Customers always gets 3–5 quotations for anything they procure and negotiates with the lowest price bidder and tries to reduce the price further. We may think what difference is there in just 1 Re., but when customer works out his volume of purchase, even half rupee shall be important. The customer uses several tons of yarn, and the cumulative savings for him shall be quite high. So we need to design our activities so that we can produce quality goods at lowest possible cost; this includes monitoring the wastes, monitoring the stoppages, achieving maximum possible efficiency, avoiding non-value adding activities, reducing the inventories, responding fast in case of problems, utilising the available resources fully for productive works and so on. Therefore, quality of work includes doing the work with at least possible expenses, while achieving the quality, productivity and delivery in time.

3.4   Aligning work to meet customer needs

We do innumerable works in an organization and it is very difficult to list them all. Whatever may be the work, we need to question in what way it is contributing to achieve customer satisfaction or reducing the cost of overall operations or improving the performance of the company. If an activity is not contributing to any of the above, we need to review and question the necessity of doing that work.

3.4.1   Planning the activities keeping customer's requirement in focus

While planning an activity, we need to keep customer requirement in focus. A customer wants to produce socks for football players in view, and then he needs a yarn which can withstand the abrasion. A combed yarn with slightly higher twist than normal hosiery yarns can be preferred. However, as the use is very rough, and the socks are to be washed very often, the life of socks is less. The white socks become dirty very fast, but coloured socks are not preferred for that game. The players are normally youth, and hence the customer wants the yarn to be cheaper. So normally carded yarn with slightly higher mixing for that count is preferred than a combed yarn. The smooth feel got by combed yarn is not considered important as the youth playing football have rough skin and the smoothness provided by combed yarn is not felt as a need. However, if we are catering to a socks manufacturer producing baby socks, then combed yarn gets the preference and the higher price quoted also is considered as affordable compared to producing carded yarn socks and using for tender babies.

As discussed earlier, the delivery in time is also a very important requirement of the customer. We need to make a combination of machinery available with us and allocate them in such a way that we can meet the dead line. We need to make our machines flexible.

If customer's requirement is in small quantities but different varieties, then we need to plan the activities on slow speed machines rather than on high speed or super-high speed machines. Sometimes use of sample machines for production of such orders shall be economical. For example, in case of dyed yarn orders, the customer requires five different shades, one with 400 kg, another with 200 kg, the third with 20 kg and the fourth with just 2 kg, and all have to be supplied together. Further customer does not accept two lots in the same shade as there may be lot to lot variation in shade. So we prefer to have dyeing machines with capacities of 1 kg, 2 kg, 5 kg, 10 kg, 20 kg, 50 kg, 100 kg, 200 kg, 400 kg and machines coupled to dye even 600 kg or 800 kg in a lot. Similar is the case with spinning also. By having all long ring frames with 1200 spindles, it may not be possible to cater to small orders or special qualities. If the management strategy is to accept any quantity of order and any count the customer asks, we need to develop flexibility in machines as well as in systems. If our customers are large quantity buyers and are more particular in cost, high speed and automated machines may be the choice.

3.5   Work quality in marketing

Marketing is a very important process in any industry which involves educating the people on the merits of the products being manufactured and creating interest in them to purchase the product, providing the products to the customers as per the agreed terms and specifications and collecting the amount from customers in time. The quality of marketing is measured by the uninterrupted sales of the product being manufactured, timely recovery of money from customers, repeated orders for the products, establishment of brand image by virtue of quality of products and timely supplies.

Marketing job is not an easy job. You need to approach customers, have to travel a lot without knowing the actual place, you may get food or may not, you need to wait in front of customers office to get an appointment, hear all the abuses he gives especially if someone has not behaved properly with that customer somewhere at some time, convince him to purchase your materials for which he is not ready, make him accept a price that is workable for you, collecting the complete requirements, following up with production people to supply the materials in time and recover money from the customer. It requires a lot of patience, proactiveness, decision making on the spot, politeness, a sweet tongue, apart from the knowledge about the product and the mode of using it.

A marketing person should be polite and pleasing while approaching a customer, an authority on the product being sold while explaining the product, quick in understanding the customer needs and take a decision of modifications to be done on the product, patient while listening to customers vows, confident while suggesting solutions to customers, precise while fixing specifications of the products to be supplied, quick in working out the costs by adopting marginal techniques and activity-based cost techniques, firm while accepting the delivery dates and fixing price, punctual in replying to customers enquiries and following up for deliveries, shrewd in collecting money from customer in time and respecting the time of customer and adjusting his time table to meet the customers time requirements.

Number of youngsters wants to join marketing as they are paid well and can visit different parts of world, but unless they have the required attitude they are going to fail. Studies have shown that rather than producing a good product, ability to sell the product plays an important role in the survival of the industry. With the advances in technology, normally all mills produce good quality, but all are not having good marketing approach, because of which they fail. A good marketing man knows where to sell, how to sell, how much to sell and whom to sell.

3.5.1   Customer communications

Communicating with customers is one of the most important activities in marketing, which is considered more crucial even when compared to selling the products and getting the payments. The communications if had properly can lead to customer satisfaction and makes him purchase our products. The customer-related communications may be handling enquiries, providing information on the status of materials in process, responding to customer complaints and feedbacks and social communication to build relationship.

3.5.1.1   Enquiries and response time

Customers shall have different types of enquiries, may or may not be related to the products and services that are being procured. The enquiries include the range of products being offered by the company, the alternative products that can be used if our product fails, the salient features of the products offered by us against the products offered by our customer, the places of interest around our factory so that the customer can spend some time, details of train and air connections, the hotels where they can stay, the items they can purchase and so on. Irrespective of the type of the enquiry, the marketing person should respond fast and provide the information to customer so that a repo can be developed with customers that can help in future for marketing the materials. A customer normally prefers the company where the speed of response he gets for his enquiries is high. The culture of an organization is reflected by the speed of response not only for the customers, but for any enquiry from anyone as one is not really sure who shall be the potential customers. The marketing personnel should develop the attitude and habit of responding fast to all enquiries. It is now a normal practice to provide hot lines to the marketing personnel and allocating specific persons for specific customers so that they can get quick response.

While responding, it is necessary to be specific in giving the information and not providing other information that are not asked for. If you are straight in giving the reply to the points enquired, the customers shall be happy and shall continue business with you. If your answers are not focussed, the customer shall doubt your intention.

3.5.1.2   Providing information on status of material in process

Customers after placing an order shall be more interested in getting the correct information on the status of materials in process so that they can plan their day-to-day activities. The customers are interested in actual facts, whereas it is seen that in number of cases, the marketing persons give false assurances to keep the customer happy. By giving false assurances, the customer gets fed up and he may start searching for another supplier who is frank in telling the exact position. Please remember that the customer is also running an organization and knows the practical problems in manufacturing and logistics. If he gets correct information, he can also help you in taking the correct action. If you are giving wrong information and false assurance, he will lose confidence in your company and shall not come back to you. Number of customers are lost not because of poor quality of materials, but because of the attitude of marketing personnel.

With the increasing competition and short cycle times demanded in manufacturing, precise information on the status of materials is very important, and people are adopting supply chain softwares for monitoring the process flow.

3.5.1.3   Responding to customer complaints and feedbacks

Respecting and responding to customer feedback is a very good indicator of work quality of a marketing person. One should understand that no customer is interested in making a complaint, but is interested in running a business. He wants smooth working at his place so that he can peacefully think and plan for future. A customer does not want to waste his time and energy in making a complaint if he is not having a confidence that the company can improve and deliver the goods and services to his satisfaction. Searching a new supplier is easy compared to educating a company which is not ready to listen to customer's complaints and feedbacks. A customer shall be happy to work with us if we are respecting his words and work towards fulfilling his needs.

In a number of cases it is seen that the top management as well as the staff in textile mills try to argue and prove that they are correct and customer is wrong in order to avoid paying claims. They give cooked readings to convince that the product supplied was really good, forgetting that the customer is using the materials day and night and can judge a product much better than the producer who tests only a sample to conclude whether the product was good or not. With this they lose the customer.

3.5.1.4   Communications to build relationship

Customer communication is not only aimed at making business for today, but to build a long-term relationship. A good relation makes customers to come back to us even in spite of some quality issues here and there. Good communication helps in building personal rapport. However, one should be careful to ensure that the time of customer is not wasted or the customer is contacted in a wrong time when he was busy with some other party or dealing with some of his personal problems. Keeping his presence of mind is always important while communicating with customer, whatever might be the purpose of communication.

3.5.2   Merchandising

Merchandising is defined as buying and selling of goods for the purpose of making profit. It is concerned with all the activities necessary to provide a customer with the merchandise they want to buy when and where they want it and at price they can afford and are willing to pay. This involves making buying plans, understanding the customer, selecting the merchandise and promoting and selling the goods to the consumer. Merchandising is a very important function for the products that are sold directly to end users like garments, finished fabrics, sarees, towels, socks, curtains, furnishing cloth, and so on.

Merchandising department of a manufacturing unit is responsible for key activities that convert product into desired volume and exerts enormous impact on a successful merchandise control. The merchandising unit act as a bridge between the production activities and buyer's expectations. A buyer expects its product to be delivered in the right time, at the right price, at the right place and at the right quality.

A merchandising department is the coordinator of all the activities at the manufacturer's end. It interacts with the buyer to get orders. It does costing of the garment with the knowledge of the product considering the desired margins and negotiates with the buyer on the price points keeping in mind the competitor's price. It takes care of product development, sampling, costing, negotiations, delivery schedules, production planning, fabric and trims orders and regular follow up.

3.5.2.1   Visual merchandising

A successful retailing business requires that a distinct and consistent image be created in the customer's mind that permeates all product and service offerings. Visual merchandising can create a positive customer image that leads to successful sales. It not only communicates the store's image, but also reinforces advertising efforts and encourages impulse buying by the customers. Visual merchandising is everything the customer sees, both interior and exterior, that creates positive image of the business and results in attention, interest, desire, and action on the part of the customer. The interest in which the displays are arranged contributes maximum for the sale of the products. A whole-hearted involvement in arranging the materials in stores and displaying them, replacing the items displayed from time to time, adjusting the light depending on the materials displayed, indicate the work quality of the merchandiser and his success depends more on his work quality.

3.5.2.2   Maintenance and housekeeping

Maintaining the shop in a good condition is very important for the success of a shop. Broken tables or chairs, broken windows, torn floor mats, faded wall paints, faded display boards, burnt tube lights, noise making fans are negative factors which bring down the sales in spite of the materials sold are of good quality. The shop manager should constantly watch and arrange for replacing the defective materials at the earliest. Similarly a good housekeeping is very important. If the store is dirty, materials thrown here and there and the sales persons not lifting the materials and walking on the same shall impact on the customer and customer shall not enter that shop again.

3.5.3   Servicing the customers

Servicing the customer includes various activities like answering to all their queries, putting all the selected garment or fabric pieces in a good bag and keeping it in the vehicle in case the customer has one, or sometimes arranging to deliver the materials at the doorsteps of the customer. Depending on the time of business one has to offer water, cool drinks, coffee or tea to the customer.

In some cases, when customers have come with their small children, additional care is to be taken to keep the children busy so that elders can concentrate on selecting the materials of their choice. The children may create nuisance but we need to tolerate. More care taken to please a customer results in customer delight and he shall be our regular customer. The sales persons should be so flexible while servicing the customers. Unless you love your job, you will not be able to do all those.

3.5.4   Handling customers in retail show rooms

In a textile or garment show room, where the materials are sold in retail, it is a normal practice that a customer would like to see 10–30 pieces for selecting one piece. The sales person should have patience and show the pieces asked by the customer with a smiling face. He need to take back the materials and refold them and put it in a new bag again. Whatever may be the quality of the product, the way in which it is presented to customer becomes a deciding point on which customer purchases the fabric or garment. While selecting the material, the customers shall be asking number of questions, which may be relevant or not for that product, but the sales person needs to answer to all of them with a smiling face. Some of the comments may be irritating even then the salesman should not lose his temper, but reply to customer in a sweet joking way, without hurting the feelings. Unless a sales person loves his job, he cannot develop this attitude.

The customer expects immediate attention to him the moment he or she enters a show room. A guide should always be present and welcome the customers with a smiling face and enquire about the product they are interested and take them to the concerned sales person. It is a monotonous job, but should be done with a smiling face.

Team working and problem solving

4.1   Concept of teams

Man is a social animal and lives in society. He takes the help of society in number of his activities and also works with society. Man cannot survive alone. He needs someone with him to share his joys and sorrows, to share his feelings, to support and encourage him, to correct and guide him, to educate him, to give courage and so on. One may claim that he has achieved a goal single handed, for example winning a race, fighting an enemy, convincing a terrorist to become a good citizen, etc. If we analyse the things in detail, we find that someone was with him. For running a race, you should practice. You will be busy in practicing. It is possible when someone else like mother, sister, brother, wife, friend, father is looking after your other needs like providing you food in time, providing the clothes that are washed and clean, not disturbing you to do works for the home or earn money for daily living. Even if we take the case of a terrorist getting convinced to become a good citizen, there are number of people playing behind the screen. Among the people dead or injured, there may be people near to him, he might have lost all his friends in an encounter and become lonely and waiting for someone to console him, rejection of him by the society members making him to accept the words when a social reformer is offering to counsel, and so on.

4.1.1   Purpose of team

Normally, we refer to teams while doing some specific project; whereas we need good cohesion among employees within a section or between sections of an organization in order to make that organization strong and competitive. An organization cannot survive if there is no cohesion between the people working for it.

     The purpose of teams is to bring people together, make them understand the common goals and put their collective efforts to achieve the goals. Unity has strength, whereas a single cannot achieve. Teams can do wonders.

We need to resolve the differences we have among ourselves. It does not mean that we are fighting among ourselves when we say we have differences. We both have the same intention that we should become success, but the path we feel as suitable is different. I know a path, in which I have confidence, but you know another, which also lead to same destination, and you are more comfortable with it. I insist on my path, where as you say that your path is convenient. We get the difference. Because of the differences, our works are not streamlined, and we lose time in streamlining the activities. The efficiency shall be less and we finally lose. Because of differences in thinking, we do not come to consensus and the works shall start late. We might put unnecessary restrictions by which the efficiency shall be lost.

It is seen that people working in different sections of the same organization remain isolated, although within the section they are all united. In addition the hierarchy separates the people, and we see number of isolated islands.

When people are clear that both are working for the same cause, and both are complement to each other, they start respecting the views of others. The most important part of a team is to understand and respect the views of colleagues, rather than insisting on our own ideas and systems. Insisting on our ideas shall make the gap wider.

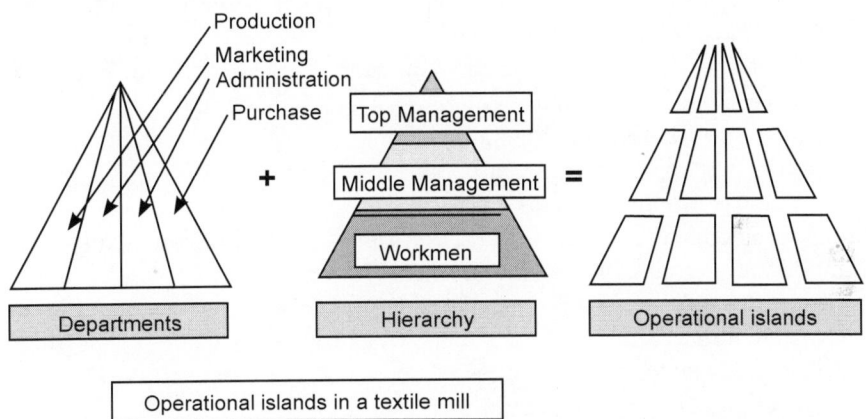

Operational islands in a textile mill

In the textile and apparel industry, certain works need to be done in groups only, whereas some are allotted to individuals; however, they also have a relation with some other activities that are done. The group activities like doffing a speed frame, doffing a ring frame, cleaning and maintenance activities, beam gaiting on looms, number of wet processing activities, material handling, stitching activities in garment manufacturing, etc., demands synchronization of works between the team members and cooperation between members. The supervisor has to ensure proper coordination between the activities of team members and also harmony in working. The success of supervisor depends more on his skills in coordinating between the team members rather than his technical expertise in setting a machine for various parameters.

The teams may be for achieving a fixed task or to work continuously in harmony for a long time. The short-term teams are normally formed to achieve certain tasks like preparing a curricula for certain training, erecting and commissioning a new machine, analysing a complaint and suggesting corrective and preventive measures, taking improvement projects for reducing wastes or improving quality or improving productivity, and so on. There are certain teams that have a long-term goal of working in harmony and cooperation among members in monitoring certain activities like safety, welfare, and ensuring compliance to certain requirements of trade and social compliances. The examples of such long-term committees are quality circles, work committees and joint committees.

The teams may be formal or informal teams. The formal teams are formed as per certain norms of the company or the government. The examples of formal teams are joint committees, work committees, welfare committees and safety committees, etc. They function as per the guidelines given by either the government or by trade. The activities involve verifying the adherence to safety norms, human rights, grievance redress, welfare measures, canteen management, welfare activities, and social compliance and so on. There are certain formal teams, normally referred as cross-functional teams, where members from different sections are nominated by the top management to accomplish certain tasks within a certain time period. It may be installing a new plant, shifting machines from one site to another, attacking certain chronic problems, preparing curricula for training modules, conducting sales promotion programme or customer meets.

The informal teams are formed either for performing certain tasks or with an intention that people in a team work together as one all the time. The examples are certain committees formed for arranging sports activities, cultural activities and quality circles.

A team is more than just a work group; it is a collection of individuals who work together. In a work group, each member is directed by and reports to a common manager or supervisor, but members don't necessarily collaborate with each other to complete their tasks. The manager of a work group usually has deciding authority. A team, by contrast, comprises individuals with complementary skills committed to a shared purpose, common performance goals, and an approach for which they hold themselves collectively accountable. The members interact with each other and with the team leader to achieve their shared goal. In a team, members depend on one another's input to perform their own work and look to their leader to identify and provide needed resources, coaching, and a connection to the rest of the organization. A team makes decisions that reflect the know-how and expertise of many people, not just the leader.

Organizations form different types of teams for different purposes. Here are some examples.

Team type	Purpose	Example
Self-directed work team	Meets are on-going and may be on daily basis to perform a whole work process	At a cotton spinning mill, a team of four people is responsible for ensuring that raw materials are purchased correctly according to company guidelines.
Project team	Gathers to address a specific problem or opportunity and then disbands. They need not meet daily, but as decided by the team	(1) Several unit heads in a garment factory explore the potential benefits of adopting a new technology, present their findings to executives, and disband. (2) A team taking the assignment of installing a boiler and commissioning.
Virtual team	Brings geographically separate individuals together around specific tasks	A project manager for garment brand hires agents from around the world to work with garment factories on a major style, while making all transactions on Internet and ERP.
Quality circle	Works on specific quality, productivity and service problems	(1) Customer-service employees and managers in a process house generate and implement ideas for improving service to the company's biggest customers. (2) A team of loaders work together and design a new method of unloading and loading.

4.1.2   Characteristics of an effective team

Whatever may the type of team or the task given to them, the team should perform. The word PERFORM has following letters, which are the characteristics of a good team.

- Purpose
- Empowerment
- Relationship and communication
- Flexibility
- Optimal productivity
- Recognition and appreciation
- Morale

It means the team should perform; if it is not performing, it is not a team but just a mob.

Purpose

Everyone has a purpose and works to fulfil it. The organization has a purpose and in the same way any team has a purpose. The team leader needs to ensure that purposes are fulfilled. Therefore the leader and the members should be clear about the purpose. To achieve the purpose, the activity of the team shall be decided and each one has to be given certain objectives and goals. The leader should align company objectives and goals with individual objectives and goals. The leader should convince the team members that by achieving company/team goals, one can help achieving personal goals.

In the process of understanding the purpose, following steps may be considered.

- Enquire your boss why you should do this work
- Understand why you are doing this work
- Explain to your people why they need to do this work
- Understand from them why they are doing this work

Empowerment

Anyone can work provided he is empowered suitably to perform the task assigned. They should not hesitate while doing a work with a confusion of whether to do it or not, whether I shall be questioned if I do the job and so on. Therefore empowering the people to do their jobs with delegating adequate authorities to perform the jobs is very important. Please remember, as a team leader if you do not empower your people, then you have to do all their jobs. When people are empowered, they shall do the work with confidence and interest that help in improving efficiency and productivity. Mind works faster when there is no tension, and efficiency increases when there is no force.

Relationship and communication

There is no meaning in staying together without good relation. When there is a good cohesive relation between team members, they work together. One has to remember the saying "Together we win; Divided we lose".

- Trust is the driving force for relationship. The leader should:
- Have trust on the members
- Understand the feelings of the members
- Communicate with members effectively
- Work together with the team members
- Be flexible; do not be rigid all the time. The rigidity may be detrimental to relationship
- Concentrate on results

Mode of communication and the methodology used has a say on the relationship between members of a team. Written communication gives clarity to the members, but should be used only for certain items where remembering is difficult. For example, when a trail is decided, the process parameters may be given in writing. While trying to understand certain act of a member, a written communication shall lead to displeasure, whereas an informal talk shall be beneficial. Sometimes phone calls may not be helpful as the members might be busy in certain activities and not able to take the phone, and in such cases a SMS or a slip of paper indicating the message in simple words may help. Sometimes, the written message might not be conveying the correct intention of the message and might lead to misunderstanding; in such cases, it is better to talk personally and convey the correct message.

Flexibility

The important point in a team is that it should perform as one, and no individual should be given undue weightage. The members have to observe certain code of conduct and the leader should be rigid in enforcing the discipline and ensuring that the code of conduct is respected. However, while working in a team comprising of people with different ideologies, if it has to perform as one, some rigidity may have to be eliminated, but the focus on achieving the goal shall remain intact. The leader should decide on how much he should be flexible with his members and in the procedures to be followed so that there is no negative feeling among the members and they work as one to achieve the objectives.

Optimal productivity

Achieving optimal productivity is one of the tasks of any team, and it should focus on the same. What is required by the team and for the company is to be understood by the team members and the leader. They should be clear as to how much is required to be produced and has to work for achieving the results. The leader along with the team members should review and find the reasons for non-achievement. The leader should facilitate the team members to achieve optimal productivity.

Recognition and appreciation

Recognition improves efficiency and brings cohesion in the team. If you do not recognize in time, people get de-motivated. The interest in the team as well as in the work reduces when recognition is not given for their efforts. In a number of cases, the management recognizes the people only when they get the results, whereas the results are a function of number of external factors. Let us take the example of a marketing team. Due to adverse market conditions, the mills might not get required price for yarn although the marketing team has done good efforts. Alternately, the quality of the yarn may be bad and because of that the marketing man is not able to push that yarn in the market. The marketing might have been successful provided the members had freedom to negotiate the price, whereas the price is fixed by top and the marketing team cannot book the order with lesser price even by 1%. Therefore, one has to see the process and recognise the people for the efforts done and not on the results achieved. Then only a team can remain cohesive.

Recognition should be done in time and not after a long time. People expect someone to recognize them after fulfilling certain tasks or after completing certain process. The recognition need not be in monitory terms, but a word of appreciation, a word of support, or a word of consoling can do a big trick.

Morale

If the morale of the team members is high, everyone feels that he/she is a part of the team. They are enthusiastic in achieving the goal and shall be really happy when goals are achieved. They will not think anything against the interest of the team. When morale is good, the members will not get disappointed with small failures as they strongly believe in their team and the efforts made and are confident of success in the next attempt. A leader should always ensure that the morale of the members is not allowed to drop down.

4.1.3   Team-building techniques

There are four stages in team building. When a team is formed, the members are new and the leader has no clarity on the competency of each member. Each member is enthusiastic as he is involved in the team and the morale of the team is high. This stage is called orientation stage. During this stage, the members introduce themselves to other members and also share their previous experiences, achievements and assure their support to other members for fulfilling the task of the team. The leader allots different tasks to members considering their capabilities as described during orientation. When the work starts the problems come to surface. The members have to face various difficult situations depending on the nature of activity and start blaming other team members for the failures in accomplishing the task. Some dissatisfaction creeps in the minds of the members and the morale comes down. This stage is called dissatisfaction stage, and the role of leader becomes very important.

Any team formed to accomplish a challenging task has to pass through this stage of dissatisfaction. If a team claim that there was no dissatisfaction, it normally means that the work was not started or the target taken was too small and not challenging. Leader needs to understand the problems of each member and work out a strategy to overcome that problem. He may have to reallocate the activities or combine certain activities or even eliminate them. He might have to take some new members depending on their skills and capabilities and remove those who cannot contribute for the task undertaken. This stage is called resolution stage. Once the resolution is proper, the team starts working and gets the production. The working of the team shall be smooth and each member starts contributing as per his capacity. This stage is called production stage. The morale of members increases as the production starts increasing.

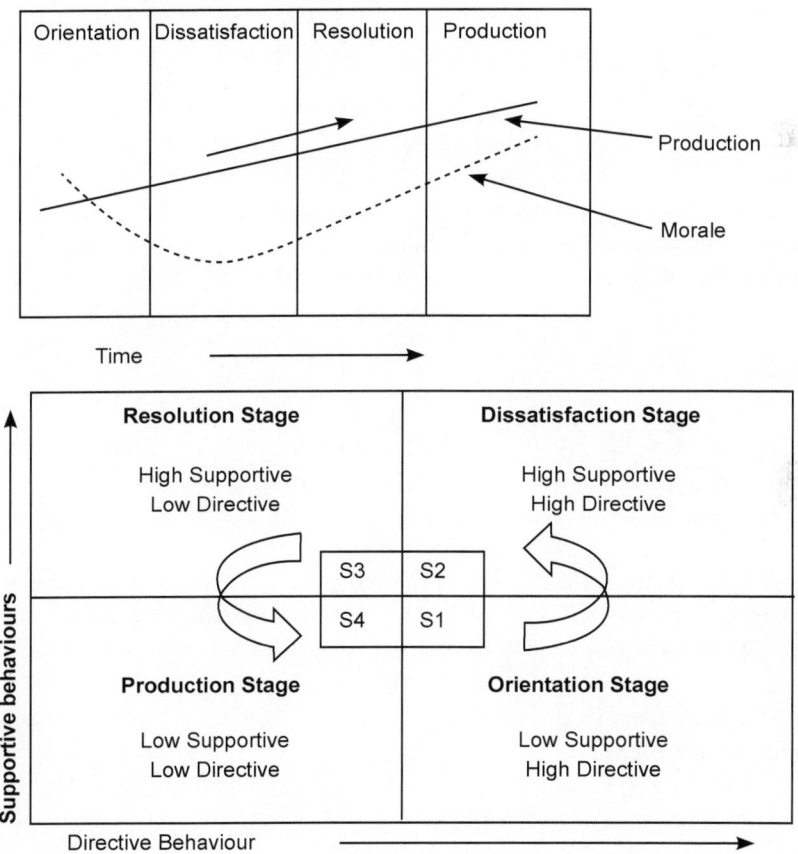

Stages of team building

| Orientation | Dissatisfaction | Resolution | Production |

Production

Morale

Time

Resolution Stage

High Supportive
Low Directive

Dissatisfaction Stage

High Supportive
High Directive

| S3 | S2 |
| S4 | S1 |

Production Stage

Low Supportive
Low Directive

Orientation Stage

Low Supportive
High Directive

Supportive behaviours

Directive Behaviour

Leadership behaviours

The team leader has to exercise different styles of leadership depending on the situation. He shall be low supportive and high directive in the orientation stage, whereas high supportive and high directive in the dissatisfaction stage. When the team reaches resolution stage, the leader shall be high supportive and low directive, whereas when the production stage is reached, he shall be low supportive and low directive. The team members will perform by themselves and there shall be no need of additional efforts by the leader. The leader shall have time to plan the further activities.

The success of a team depends on the leader. He should be in the team along with the members, holding the team members in their respective positions as a thread in a garland, which is not visible from outside, but holding all the flowers in their respective position to make the garland beautiful. The leader should be motivating the team members to get the results on a continual basis.

4.2   Quality circles and cross-functional teams

Two types of teams are popular in making improvement efforts, viz. quality circles and cross-functional teams. Quality circles are team of people from a section working in a similar environment doing identical jobs and coming voluntarily together to solve some of their work-related problems. The cross-functional teams are nominated by management taking experts from different sections to perform a job or to attack a problem.

4.2.1   Concepts of quality circles

The concepts of quality circles were developed and started in Japan in 1950, by Dr Ishikawa. The very important aspect is creating an environment by the top management that workers feel the importance of voluntarily coming forward to solve their work-related problems. The faith in management by the workmen is the driving force, and not any financial incentives.

The transparency in the top management actions and policies shall influence the people to come voluntarily forward to take some of the load on themselves and work for solving the problem. We cannot form quality circles by force.

The main idea of a quality circle is to make people work as a team and not solving the problem.

Normally a quality circle consists of 5 to 9 people, and they meet periodically and discus the problems and initiate corrective actions. Once they get the results, the same is presented to the top management, who recognizes the efforts and rewards the teams suitably so as to increase their morale.

The quality circles are self-evolved voluntary teams who identify the problems themselves and try to solve them. This cannot solve bulk of the problems that are either technical related or system related, which needs specialist teams to work for that specific task. Normally 80% of the problems need management interference and 20% can be handled by quality circles. If a problem is found as important and the quality circles have not identified that as a problem, the management shall have to find a way to solve them.

4.2.2   Concepts of cross-functional teams

Any problem shall have its roots spread at different places, and hence to solve such problems, we need to take the help of concerned people. Hence a cross-functional team is needed. The management identifies the problem and nominates these teams. The team members shall be assigned specific authorities and responsibilities. The team members in a project team shall be experts in their area and by a combination of knowledge and experiences, along with their authoritative positions, the problems shall be solved.

The cross-functional teams or project teams shall have a time limit to complete their jobs, and after that time is over, whatever might be the position, the team shall be dissolved and a fresh team is formed to solve the remaining part of the problem.

The teams follow various steps of problem solving, viz. problem identification, observation, root cause analysis, devising alternate actions, taking suitable action, verifying the results and reinforcing the systems.

4.3   Problem solving

People normally discuss about problems. There is no one without a problem. Each one has a problem or a number of problems. All might not be having same problem. If I describe a problem, another might say it as "not at all a problem". Something is problem to me, but the same is not a problem to you. Then what is a problem? There are different definitions. According to Oxford Dictionary, it is a doubtful or difficult matter requiring a solution. It is also defined as something hard to understand or to deal with. A problem is defined as an undesirable result of a job as per the Japanese concept of total quality management. The solution of a problem is to improve the poor result to a reasonable level. If we solve our problem, we can move forward. If we leave it, it shall go on growing.

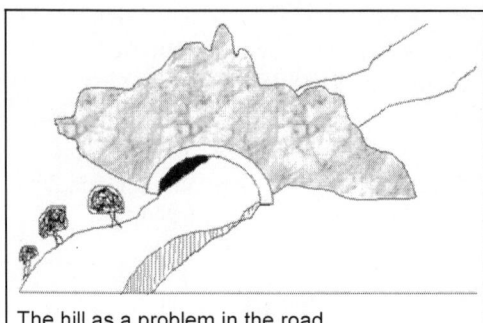

The hill as a problem in the road

In the picture we can see a hill, which was obstructing the road. People had to take a long route because of the hill. Once the tunnel is made, the problem is solved. Vehicles can easily pass through, saving time and fuel. Therefore solving a problem leads to improvement. Each problem should therefore be considered as an opportunity for improvement, and we take suitable actions to overcome the problem.

4.3.1   Identification of problems

Take the example of a tree in my yard, which I do not want. It is a problem to me, as it sheds lot of leaves, and house is getting cracked because of its roots. I have tried to cut it off, but still it is growing. I am not putting any water, but it is managing by taking water from the neighbours. What can I do? How to solve the problem?

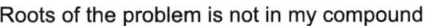

Roots of the problem is not in my compound

When Dr Juran told that 80% problems were management related, people laughed at him and branded him as mad. Let us see my example. I have a tree which is a problem to me. Who is responsible for that? I did not remove it when it was a small plant. I allowed it to grow. The roots spread. I am not putting water, but it is getting. Are my neighbours responsible? No. They are not aware that roots are there in their yard. Why the problem came? The land

and soil in my yard is suitable for that tree to grow. My systems are favourable for the problem to grow. I should have reinforced my land by putting stones or cement, so that a plant cannot grow. Now what I have to do? I should take the neighbours into confidence, and dig the roots out from their yard. Once all roots are removed, I have to put stone pavement to prevent a second plant from growing in my yard. This holds well for all the problems and for all organizations.

4.3.2   Seven steps for problem solving

Seven steps for problem solving designed by Dr Juran and Prof Ishikawa are relevant for all. They are as follows.

1. *Problem identification* – Define the problem clearly
2. *Observation* – Recognition of the features of the problem
3. *Analysis* – Finding out the main causes
4. *Action* – Action to eliminate the causes
5. *Check* – Confirmation of the effectiveness of the action.
6. *Standardization* – Permanent elimination of the causes.
7. *Conclusion* – Review of the activities and planning for future work

Step 1: The first step, defining the problem, consists of highlighting the importance of the problem and the need to solve it, the background of the problem and the course of actions taken so far, the present loss because of the problem and the extent to which the problem is required to be solved during the present task, and the budget sanctioned for solving this problem. Use of as much data as possible is essential to identify the most important problem. When a problem is selected, one must be sure of reasons for selection. The circumstance in which the problem gets priority is to be identified and highlighted. The undesirable results of the poor performance are to be expressed in concrete terms, especially in terms of money lost on an annual basis. If the degree of importance is extremely high and is widely understood by many people, the problem will be dealt seriously. Next, the loss in performance in the present situation and the advantage of effecting improvements are described. The basis on which the target values are set in the theme and how it is important are indicated. When the theme includes many types of problems, then they are to be divided into sub themes for effective handling of the problem. Brainstorming is often used to identify all possible reasons for the problem to occur.

Step 2: The second step, the observation, consists of investigation of the specific features of the problem from a wide range of different viewpoints. Four different viewpoints considered are time, place, type and symptom. As the intensity and effect of problem depends on various factors, the investigation must be done from different points of view to discover variation in results. One needs to go to site and collect necessary information that is not put in the data form. The objective of this step is to discover the factors that are responsible for causing the problem.

Step 3: The third step is to find out the main cause of the problem. The activities include setting up of hypothesis, verification of all links using cause-and-effect diagram, deleting the elements which are not relevant and testing the hypothesis to identify the real root of the problem. Use of cause-and-effect diagram is made to ensure collection of all possible knowledge concerning possible causes. The information obtained in the observation step is used to delete the elements which are not clearly relevant. Then the cause-and-effect diagram is revised by marking the elements which have a high possibility of being the main causes. New plans are devised and data collected to ascertain the effect of those elements on the problem. The data are validated by small experiments, and some time the poor result is intentionally reproduced.

Step 4: The fourth step, action, is to eliminate the main causes. A strict distinction must be made between actions taken to cure phenomena (immediate remedy) and actions taken to eliminate casual factors (preventing recurrence). The ideal way of solving a problem is to prevent it from happening again by adopting remedies to eliminate the cause of the problem. While taking action care should be taken to ensure that the actions do not produce other problems. If it is inevitable, devise remedies for the side effects. The action has to be thoroughly evaluated and judged from a wide range of viewpoints as possible. It is advisable to conduct trials and check. Devise a number of different proposals for action; examine the advantages and disadvantages of each and select those which the people involved agree to. An important practical point in selecting actions is ensuring the active cooperation of all those involved.

Step 5: The next step is checking to ensure that the problem is prevented from occurring again. This involves comparing the results before and after the implementation of the solution. The data should be collected in the same format as it was done for analysing the problem, and comparison is to be made for before and after. It is also essential to relate the situations before and after for understanding. Then effects are converted into monitory terms and compared with target value. If the undesirable results continue to occur even after actions have been taken, the problem solving has failed. We need to go back to the observation step and start again.

Step 6: The step of standardization is to eliminate the cause of the problem permanently, by devising the procedures to perform the activities and documenting them. Without documenting the procedures and standardizing, the actions taken to solve the problem will gradually revert to the old ways and lead to recurrence of the problem, and also it is likely to revert when new people are on work. Standardization will not be achieved simply by documents. It must become a part of the thoughts and habits of the workers. Education and training to all involved along with assigning responsibilities is an important part of this step.

Step 7: The last step is reviewing the problem-solving procedure and planning future work. The activities involved are summing up the problem remaining, planning as to what is to be done for solving the remaining part of problem, and verifying what went well and badly. One should realize that a problem is never perfectly solved and an ideal situation almost never exists. It is not good to aim for perfection or to continue the same activities on the same theme for too long. When the original time limit is reached, delimiting the activities is important. Even if the target is not reached, a list should be made of how far the activities have progressed and what has not been attained yet. It might be worthwhile to live with a problem rather than eradicating it fully, which depends on the nature of the problem remaining and its after-effects, the steps and cost needed to eradicate it fully.

A systematic approach as mentioned in above seven steps reduces the number of problems and helps in moving towards zero. As you solve a problem, the quality of your work improves.

4.3.3   Use of QC tools

The most important step in problem solving is the identification of the real problem and its root. We need some tools for analysis and diagnosing the problem. Those tools are popular as QC tools. Seven QC tools were recognized during the evolution of TQM concepts in Japan. They are Data Collection, Check Sheets, Stratification, Brain Storming, Cause-and-Effect Diagram, Pareto Analysis and Scatter Diagram. With the time, a number of other tools were recognized and added in the list. They include Histograms, Force Field Analysis, Critical Activity Chart, Flow Chart, Concentration Diagram, Run Chart and Control Charts, Spectrograms, Boundary Analysis, Root Cause Analysis using 5-Why Technique, etc. To understand the impact of potential problems, Failure Mode Effect Analysis was developed for use at designing stage itself, which is supported by Quality Function Deployment for optimizing the process parameters. For implementing the actions, management tools were developed, which include Affinity Diagram, Relation Diagram, Tree Diagram, Matrix Data Analysis, Matrix Diagram, Process Decision Programme Chart, Arrow Diagram, etc. Let us now discuss some of the QC tools.

Data collection

Data are the building blocks on which fact-based decisions are made. The collections of facts and figures which can give a clear picture of a required work situation are called data. Data collection is the most important factor influencing the success of a problem-identification process, which is the first step in any of the improvement projects. The data would form a sound basis for decision making and corrective action. The method of planning, organizing and auditing the process of data collection are key factors, which can 'make' or 'break' any improvement effort. The sincerity of collecting data is very important to get the correct reason and results.

The primary purpose of collecting data is to answer questions, which may come from opinions during different stages of problem solving and decision making. Accordingly data are collected, for example, for the purpose of understanding the actual situation, analysis of the causes for various effects, for process control to determine whether the process is in control or not, regulating data for decision making for acceptance or rejection. It is necessary to verify the reliability and correctness of data, its relevance to the problem and ability to reveal the facts.

Data should be either measurable or countable to make analysis and study the improvement trends. There are some data, which can neither be counted nor measured like smell, taste, fastness properties of dyed material, yarn/fabric appearance, feel of a fabric, etc. It is needed to convert them into some measurable terms. Information can be obtained by careful observation of facts and analysis of data. The data can be obtained by referring to past records, actual measurements, enumeration, sampling, controlled experiments, surveys, etc. In a number of cases, data required may be available in some form, and we need to put them in a required form to facilitate analysis. Which data and how much to be collected depends on the nature of problem. The following 10 steps can be used as a guide to design a data collection system.

1. Formulate good questions which can lead to the root cause.
2. Consider appropriate data and analysis tools.
3. Define a comprehensive collection plan.
4. Anticipate bias in collecting data, and take measures to avoid or minimize.
5. Understand the data collectors and their environment.
6. Design a simple data collection form.
7. Prepare the instructions for use.
8. Test the forms and instructions.

9.   Train the data collectors to be focused on the problem.

10.   Audit the collection process and validate the results.

One need to ensure that right type of instruments are used for data collection, are calibrated and maintained. Proper definition for classification should be made for enumerating data. Factors which can influence data are also to be recorded. Data should be recorded honestly and nothing should be left out.

Brain storming

Brain storming is used to help a group to create as many ideas as possible in a short time as possible. Normally one man cannot have the complete experience or knowledge of a situation, so it is necessary to involve all concerned from various sections where the roots of the problem are spread. The subject shall be made clear and specific to the participants as it helps to focus their thoughts and ideas. Each member tells one reason at a time in rotation, and when he is not ready, shall say 'pass' and allow the next person to tell. There shall be no discussions while the points are being told, and no one will laugh or comment on the points told by others. There is no need of giving any explanation or justification while the points are being told. All points shall be recorded on a black board or a flip chart to avoid repetition. This is a very good group education technique, which eliminates bias to some extent and brings a feeling of oneness in the group or team as the participants sit together in sharing their experiences through ideas.

Flow charts – process mapping

To analyse a problem and finding solution, it is necessary to understand the process. Mapping of the process and preparing a flow chart showing sequentially the inputs, activities and processes, checking done and the controls exercised, the feedback loops, the decision points, intermediate and final outputs helps in understanding the process. The flowchart is self-explaining and does not give any interpretation by itself; however, when the ideal flowchart is compared with the actual, it shows points of deviations. A new process or plan can be tested for its logical consistency by following all paths of the flow chart.

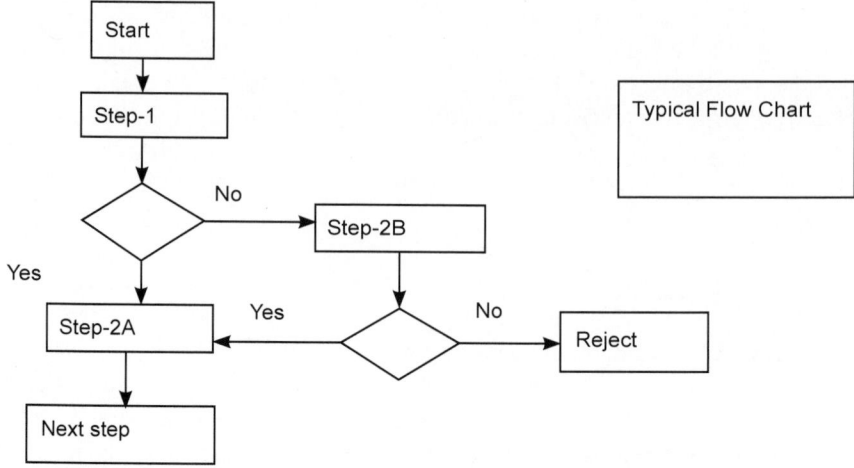

When we get a problem, it is essential to analyse the path in which the work took place and verify the reports and the actions mentioned. Identify the possible reason for this problem to remain unnoticed during the process. If we have a habit of documenting all the events that took place while carrying out an activity, we shall be able to identify the problem in its root itself and does not allow it to grow. In majority of the cases, we fail to identify the problem in the beginning because of the lack of documentation. It holds well not only for the textile mills, garment factories and other organizations, but also for the activities at home and society.

Critical activity chart

After understanding a flowchart, information is gathered on each step to understand its criticality in process. The critical activity chart is a tool for systematically gathering and analysing information about job process or operations, concentrating mainly on inputs, outputs and basic processes, rather than specific interlinked sequence of process steps. This is used to understand the work process and define its boundary in problem identification stage, to help brain storming of problems in work place and to identify major causes for dissatisfaction of internal customer, to identify the major areas of output and their internal customers and assess the extent of their satisfaction. To prepare a critical activity chart, the major work activities are identified and for each activity one chart is prepared. The next step is identifying the inputs, outputs, tasks and the customers. Then problems or deviations in output, input and processes are identified. Problems can be clarified by discussing the contents with all concerned. It is a tool for common understanding. A good Critical

Activity Chart helps in highlighting the workplace activities, and if analysed systematically, makes brain storming faster and more useful.

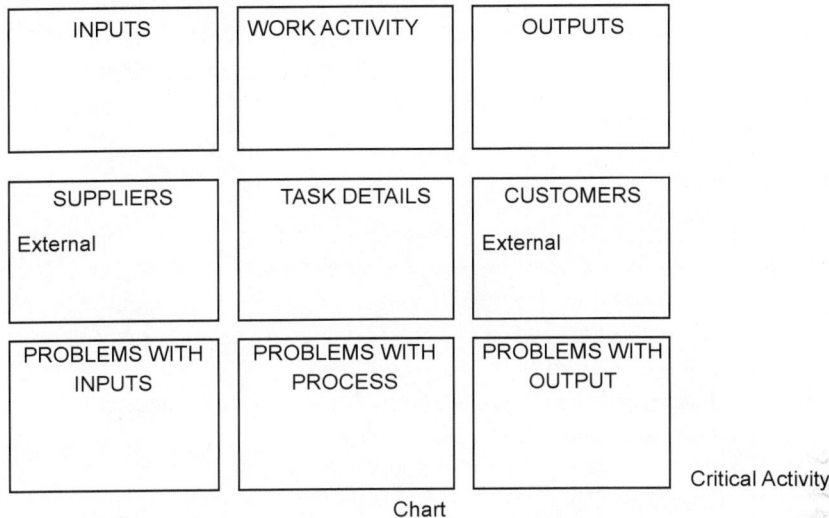

INPUTS	WORK ACTIVITY	OUTPUTS
SUPPLIERS External	TASK DETAILS	CUSTOMERS External
PROBLEMS WITH INPUTS	PROBLEMS WITH PROCESS	PROBLEMS WITH OUTPUT

Critical Activity

Chart

Boundary analysis

A boundary analysis illustrates the relationships that exist within the processes. It details all interactions among processes, customer and suppliers. First the starting and ending points of the process are defined. Then details are prepared for each process.

- Customer input → perceptions, requirements, complaints, expectations
- Process output to customer → information, deliverables (products/services)
- Supplier input → information, deliverables
- Process output to supplier → perceptions, requirements, complaints, expectations
- Resources à equipment, people, budget, procedure, training

Boundary analysis comprises of finding the boundary of a problem or its effect. For example, the tuft size if very big which was fed to a bale breaker can result in jams at bale breaker and the next machine, uneven mixing of materials till it reach the multi-mixer or auto-mixer increases in neps in the blow room material but cannot contribute for count variations. If starting marks are the problem, we need not search in warping or sizing but attend only weaving machine.

If absenteeism is high at a particular section, it might be due to poor working condition or the typical behaviour of the boss and not the HR policy of the company. If the problem is throughout the organization, then it is the HR policy and the top management and never the workers who are absenting. By analysing the nature of problem, one can point to the source of problem. There is no need to search at all places.

Check sheets

Check sheet is a well thought out format for collecting and compiling data for events as they happen which makes it easier for subsequent analysis. It may be a form, format or table. The check sheets are useful in the following areas.

    i.   To understand the past and present status of the problem situation
    ii. To stratify the data as they are collected
    iii. To understand the change through the passage of time trend
    iv. To analyse the data as they are collected
    v.   To determine the details of defects
    vi. To determine where the defects occur
    vii. To inspect machines or equipments
  viii. To verify the operating procedure

The check sheets are of three groups, viz. check sheets for recording data and making surveys, inspection and validation check sheet and check drawings. The check drawings are helpful in locating the exact location of defects to identify problem area. Concentration diagram is a type of check drawing.

The preparation of checklist involves the following steps. It starts from clarifying the objectives by clearly stating the event or issue being observed and what data pertaining to it is to be collected. Depending on the type of the problem one needs to decide as to when and where the data is to be collected and type of check sheet. Depending on the role of each item in the problem, the items to be checked are decided. First a trial check sheet is prepared to ensure its suitability for collecting the data. The check sheet should include the title, object/items to be checked, checking method, date and time of check, the checker/observer, the location and the summary of conclusions. While recording the observations, simple note using symbols can be made so that maximum information can be gathered in one stroke. The information collected is to be tallied for their completeness. Completely filled check sheet offers clearly visible data for the event and is self-explanatory.

Concentration diagram

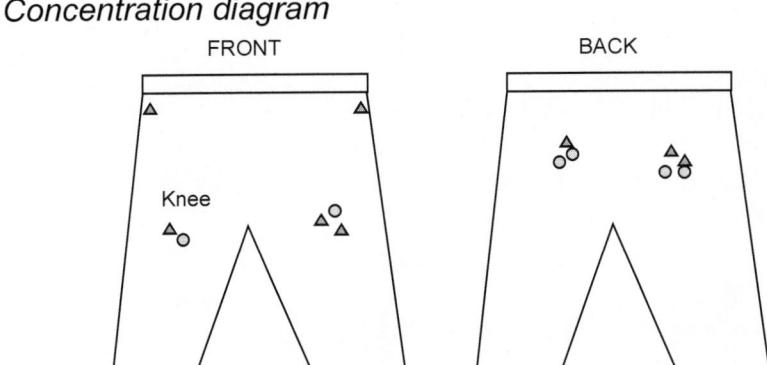

Concentration Diagram – Wearing out of pants

△ = 1 reading
O = 10 reading

Concentration diagram is a special check sheet to record data about frequency, type and location of events (defects or errors) on the picture or schematic drawings which is easily understood and visualized. It is used when visual picture or layout of location of event is more clearly understood than possible description, the possible locations are many and proper classification and expression by words are difficult, and when standard check sheets becomes difficult to understand for data collection in remote locations like the exact point of defect. The Concentration Diagrams when completely filled show the frequency and location of the event. The form is self-explanatory, as it indicates the location of the diagram. This can be further analysed by using other QC tools.

Stratification

Stratification means the separation of data into categories. This is a statistical technique of breaking down values and numbers into meaningful categories or classification to focus corrective action or to identify true causes. This is used to identify the category which contributes to the problems being tackled. Graphs are among the simplest and best techniques to analyse and display data for easy communication. Stratified data is normally displayed in Bar Charts, which show comparative characteristics by the length of the bar.

Selection of Stratification Variables is essential for planning the variables and collecting data, rather than going on adding the variables as and when some information is obtained. The cost and time for collecting additional identifying variables in the initial times is lesser compared to the cost of new collection effort. The next step is to establish category to each variable, which is a value or a range of values of a stratification variable. Then data are collected pertaining to those variables. The data obtained are sorted and grouped into stratified variables. If the first attempt of stratification does not reveal any significant pattern, data are collected again to find out the effect of other variables.

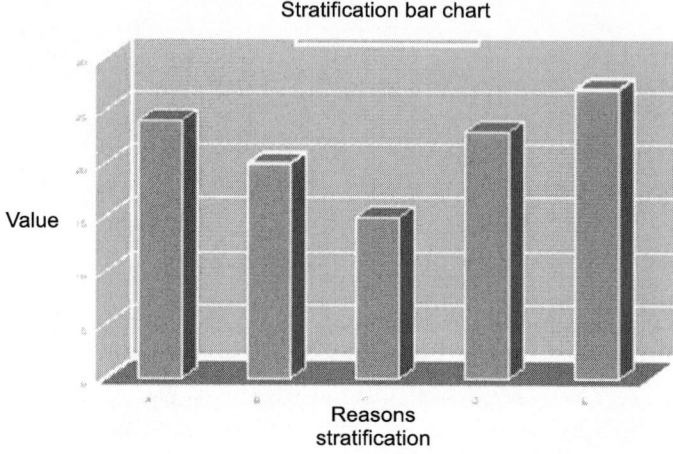

Stratification bar chart

Stratified run charts

Instead of using a bar chart, use of a run chart is adopted for analysing the system related problems. Let us take an example of a process and the time taken for doing different processes as shown below.

Time required for the process – minutes						
Process	A	B	C	D	E	F
1 March 07	20	43	25	32	25	12
2 March 07	21	40	25	35	26	14
3 March 07	20	35	25	31	24	16
4 March 07	21	43	26	30	25	17
5 March 07	23	30	25	28	26	13
6 March 07	21	26	26	42	27	15
7 March 07	20	40	25	35	25	16

For doing a job, number of steps are involved, and each step has its own importance and criticality. When we analyse the time taken for each element of a job, the normal practice is to collect data element wise and project in a Bar Chart, where either the average time or total time is shown. However, this cannot identify whether there is a real scope for improvement or standardization or a technological change is required. Normally the tallest bar is taken as a target for improvement project. When we draw a stratified run chart as shown in figure, we can identify the process which have wide variations and the processes which are consistent. The processes with variations can be improved by standardization, where as if a process is consistent, improvement can be achieved by technological change.

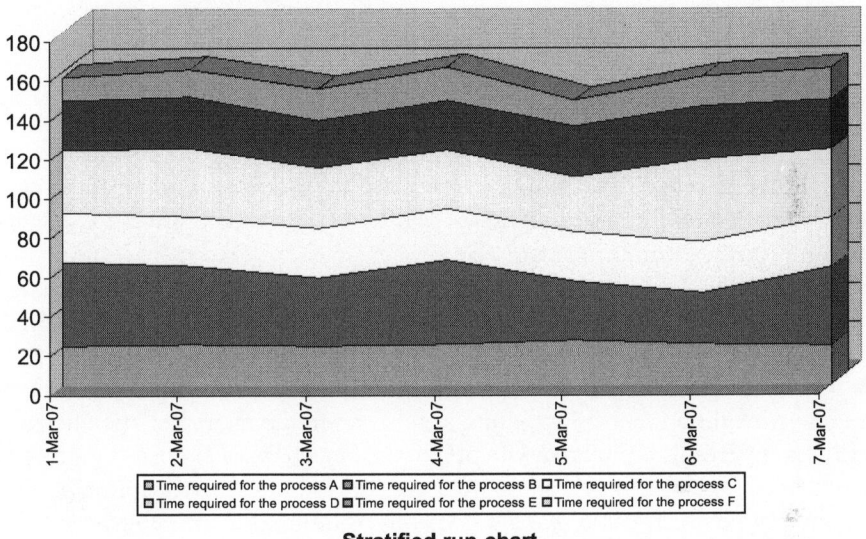

Stratified run-chart

Stratified run charts can be used for reducing the wastes, improving efficiency, implementing Lean systems, reducing delays in maintenance activities, etc.

Run charts and control charts

Run Charts display the trend of changes of a character over a period of time. The X-axis always refers the time and Y-axis indicates the character under observation.

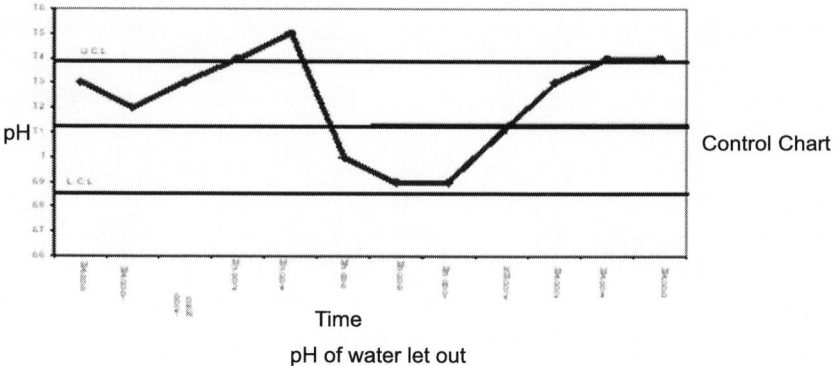

Control Chart

Time

pH of water let out

This is a specialized graph, which uses connected lines instead of bars to illustrate data. A Run Chart with statistically derived control limits is a Control Chart. This is used to highlight the variations of a characteristic over a period of time and seek explanation for changes, to study the growing or declining trend of the average, to highlight significant improvements in performance after implementing counter measures and to verify the effectiveness of the measures taken for improvement.

To make a Run Chart, time intervals are marked in the horizontal axis. The numeric scale must move in regular intervals. The vertical scale is marked considering the expected range of variation. A typical Run Chart is shown in the figure for pH of water let out from a process house in figure. Run Charts are interpreted by identifying points in time when the characteristic changes significantly. If it is compared with other possible changes at the same time, it offers clues for causes. A common trend may be increasing or decreasing. The action is to be taken if the reading goes out of limits or continuously remain near any of the control limits.

Cause-and-Effect Diagram

Cause-and-Effect Diagram is a representation of the systematic relationship between the event under investigation and all possible causes influencing. It is also a documentation of group thinking process to investigate the root cause of the event. As it looks like a skeleton of a fish, it is called as Fish Bone Diagram and also as Ishikawa Diagram in the name of its founder. This is used to investigate the cause and effect and help stratification for collection of data to confirm relationship and evolve counter measures. The steps involved define clearly the problem or effect or event for which the cause is to be identified. A horizontal line with an arrow at the right hand end and a box in front of it is drawn. Problem statement is written in the effect box. The next step is identifying the causes in major categories. Brain storming is normally used to identify the possible major causes. After identifying a primary cause, the team shall go in deep and identify as many secondary or tertiary causes as possible in each of the primary causes. Each of the major causes is placed in a box horizontal to the first line and connected to that line at an inclination of approximately 70°. After identifying the major causes, the root causes are investigated by adopting root-cause analysis techniques. The logical validity is checked for all causes identified considering the present scenario. It is important to understand the potential pit falls while using Cause-and-Effect diagram. It should not be treated as a substitute for data. It should be drawn only after preliminary data has been collected to narrow down the focus of a problem. One should not limit himself just to those theories which are in the diagram.

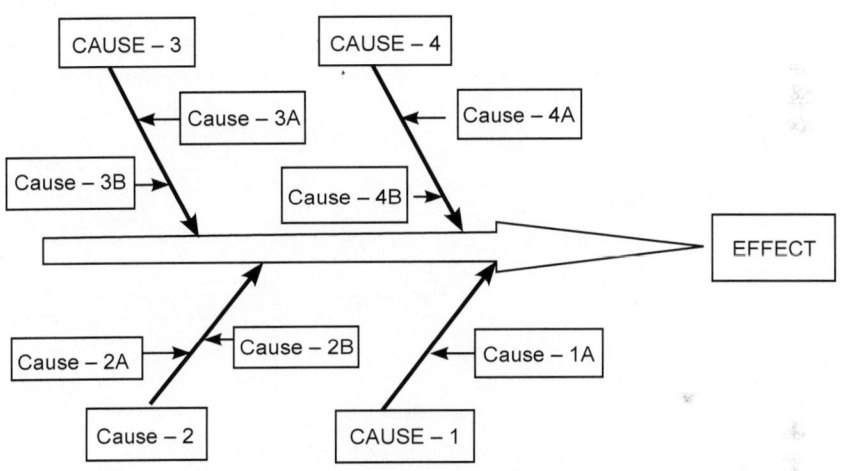

Cause-and-Effect Diagram

Pareto Analysis

A Pareto diagram is a special form of vertical bar graph which helps in identifying "vital few" from the "useful many". Its concept was given by Wilfred Pareto, an economist from Italy, and was developed as a QC tool by JM Juran. The principle involved is that very few causes contribute for maximum effect, whereas a number of other factors contribute only for a small effect. It is used while setting priority while selecting the problem, and for identifying the most important root causes contributing substantially to the problem.

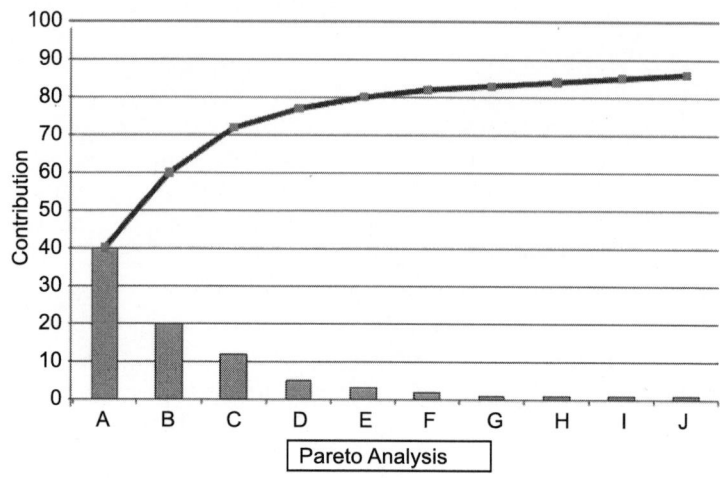

The impact of each factor is worked as a percent of total impact, and the factors are arranged in a descending order. A bar chart is prepared. In addition to a line graph is prepared for the cumulative impacts worked starting from the highest contributing factor.

Histogram

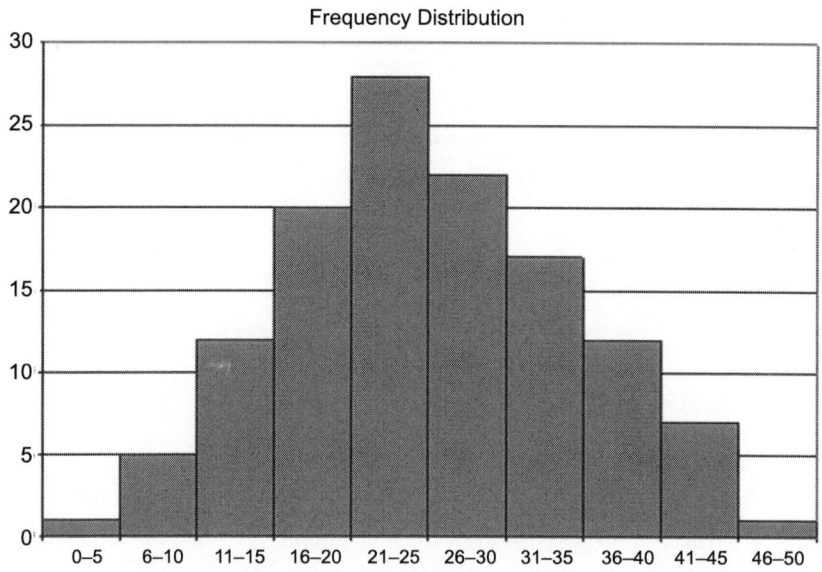

A histogram is a special type of bar chart to show the distribution or spread of the observed characteristics, which enables one to see patterns that are difficult to see in a simple table of numbers. It is a visual presentation of range, magnitude, central tendency and the spread. Histogram was first developed by French Statistician A. M. Guerry in 1833. The histogram helps to identify whether a spread is normal or not, the surprises in the natural distribution which can lead to causes or counter measures, confirms the results of a process improvement and to obtain clues for stratification. The common histogram patterns are Normal Bell-Shaped distribution, Double Peaked Distribution, Plateau Distribution, Comb Distribution, Skewed Distribution, Truncated Distribution, and Isolated Peaked Distribution or Island Distribution.

Scatter diagrams

Scatter diagram is a simple graphic presentation of the relationship between two variables, which relates to cause and its effect. It is one of the oldest applications of graph and was developed by Dr Burton in London in 1794. In 1832, JFW Horshel fitted a curve to the scatter diagram. The points marked on a scatter diagram forms a pattern which indicate the degree and nature of relationships, which is statistically known as correlation.

Scatter diagrams are used to verify if there is any relationship between cause and effects with facts and to estimate the strength and nature of the relationship between two sets of data. The concept is that there is always a relation between a cause and effect, but it would be difficult to state them in precise mathematical terms. It is easier to see the relationship in a scatter diagram than in a simple table of numbers. The effective problem solving is possible only when we discover and test the true relationship between a cause and its effect. Normally the suspected cause is taken in X axis and the effect in Y axis. Points are plotted for each reading of cause and effect. The pattern generated by the cluster of points gives clue to the possible relationship.

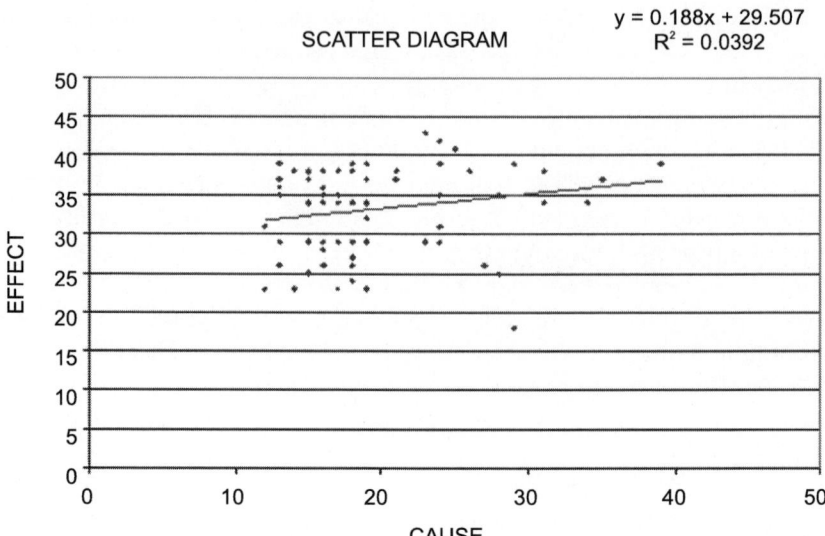

The scatter diagrams indicate the relation as strong positive, strong negative, weak positive, weak negative and no relation. There may be some complicated relations, where the value of Y increases as X increases to a certain extent, and then it might change its direction. The relation need not be linear all the time. Depending on the degree of relationship, further statistical analysis or verification can be carried out. Also quantification of degree of relationship can be carried out by working out coefficient of correlation.

The most common mistake with regards to scatter diagrams is to fail to use them, as people assume that a relation exists and need for showing it diagrammatically is not felt. Sometimes, we get correlation without physically understanding the reason or relationship.

The scatter diagram is a useful tool, but it cannot substitute the team's knowledge and fundamental understanding of the process and the problem under study.

Force Field Analysis

In the improvement process, if the improvement has to be successful, some change has to take place. There are some hindrances for the change and some elements support the change. Force field analysis is a technique developed by Kurt Lewin to identify elements which resist the change (hinder) and which are pushing for change (aids). This helps in developing the implementing strategies for a change by carefully working with the factors which favour or hinder the process. This is used for chalking out possible implementation strategy for an improvement, to forecast and assess the problems likely to occur from hindering factors while implementing a change, and to develop counter measures to minimize the impact of hindering factors during successful implementation.

If the helping factors are more powerful, the change takes place and implementation of counter measure is successful. Normally each hindering factor has a counter factor which can help. Force field analysis should be analysed to find these couples. If this is not obvious, it is possible to generate additional 'drivers' to facilitate implementation. Then if helping factors are nurtured, implementation has better chances of success of success.

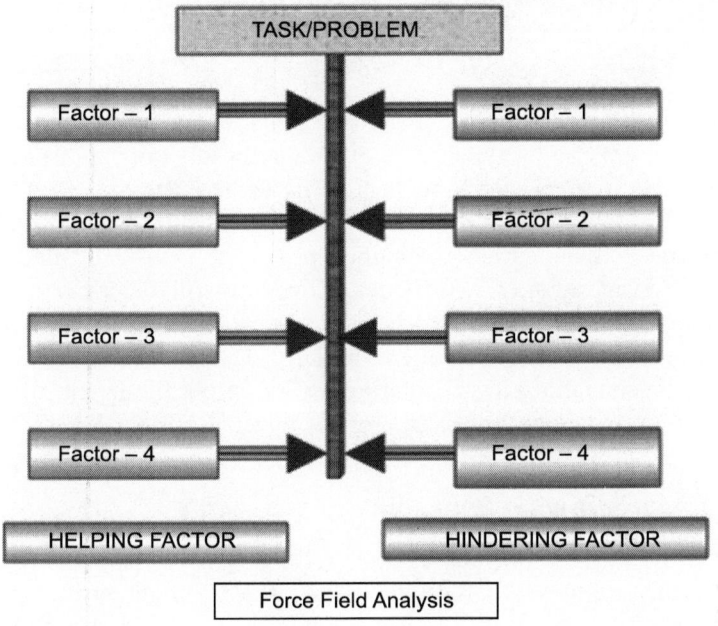

Spectrograms

Spectrogram is a QC tool being used in textile mills to locate the source of fault in a yarn, filament, rove, sliver or any such continuous strand, which are produced by using rotating rollers. It highlights the defects which occur in a regular frequency. By carefully studying the gearing diagram and working out backwards, it is possible to pinpoint the source of defect. This tool can be used in manufacturing industry producing drawn wires, where a pair of drawing roller draws the wires by applying certain draft.

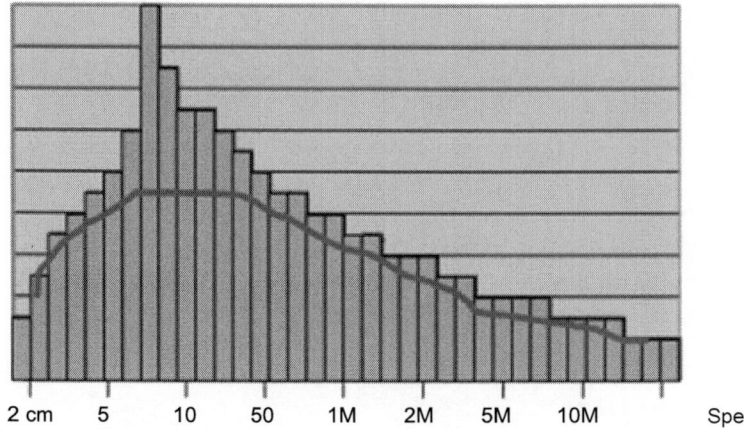

| 2 cm | 5 | 10 | 50 | 1M | 2M | 5M | 10M | Spectrogram |

The dark colour line is an ideal curve, whereas the actual values, i.e. the frequency of a particular type/size repeating are shown in bars. If the height is more than the ideal curve, then it needs to be corrected. However, the readings which are more than twice the height of ideal curve are considered as significant. In the diagram, the highest peak is at 8 cm, indicating a defect occurring at every 8 cm of the product. If the draw rollers have a diameter of 25 mm, then for every one revolution we get $25 \times 3.14 = 7.9$ cm. It means there is a problem in the draw roller. Similarly, we can work out the source of defect by understanding the gearing and working out the length of material produced for one revolution of each roller or gear.

Root-cause analysis using 5-why technique

This technique is normally used in investigating the reason for a problem by asking the question why for a number of times. For any answer received, again a 'why' is asked and finally after number of rounds we get the real reason or the root of the problem. We may get number of answers for one "why". Ask "why" for all "whys". You may get repeated reply or sometimes a reply which was contradictory to the earlier reply. Make analysis of all "why" and you will get the root cause.

This technique can be used for problem solving, enquiries for misconduct, post-mortem of accidents, breakdowns of critical machines or systems and so on.

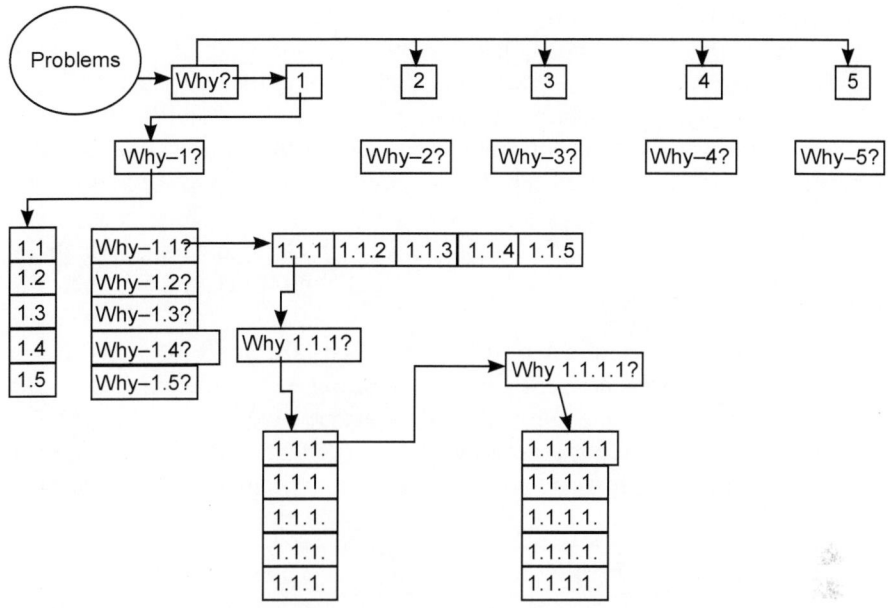

4.4   Enjoying the work by developing others

You might get a job of your liking, but sometimes you might have to accept a job that is available. Similarly you can get married to one whom you liked, but sometimes due to some reason, you might have to marry someone else. Every time you are not a chooser. You will have to accept the situation and challenges and do the work.

Working can be enjoyable when we love the work we do. When we do a work of our liking, we will not feel tired. One who loves the job will not feel tired whereas a man who does the work because of force feels tired fast. The team which likes the work will be more enthusiastic. The performance shall be good when man does a job of his choice. A leader enjoys the work when his team does the work with full enthusiasm. People get mastery in the job they like and practice.

4.4.1   Developing subordinates and co-workers

One of the biggest problems seen in number of cases in the textile and garment industry is not developing subordinates to shoulder responsibilities. It may be due to fear that subordinate might overtake him or leave the job after learning, or may be any other reason. The seniors try to maintain their works as secret.

Because of this attitude, the subordinates are leaving the job and joining other companies with a hope that they can come up.

If you love your job, you feel interested in doing the job and shall enjoy the same. You shall be enthusiastic and never feel tired. Your actions shall be replicated by your assistants, and you shall get the results faster and can emerge out as a winner. Even though you become old you will be still young. You will grow stronger and stronger. Similarly you should love your people and be one among them. You should not always think of something which did not happen and you wanted it to happen. Understand the reality and accept it. Love the present and live.

Develop your subordinates so that they can take some of your load or you will have to do all the work yourself. Subordinates shall be happy to take higher jobs that are challenging when they are confident of your support in developing them. If there is no challenges and guidance from boss, people do not like to work in such an environment. You will not get the work quality.

4.4.2   Developing love for work

Everyone does the work, face challenges, get success and sometimes failures. But a man who enjoys the work has more chances of getting success than a man who feels it as a burden. If you are grumbling while doing work, you cannot concentrate and extract the best out of situation. You cannot achieve the quality of work as required, as you are not interested, not analysing the reasons for failures, not taking corrective action, not making efforts to prevent the problems, but are always trying to find a scapegoat to pass the blames. By this, it is not only that you are failing in this accomplishment, but also are developing a habit of passing the buck and a negative attitude. When you feel your work as a burden, you shall get tired fast and your thinking comes down. You shall lose confidence when the thinking and efficiency comes down. You feel like not doing the work at all. Avoiding a work because of the fear of failure is like adopting a more painful treatment than the pain itself. If you feel the work as a burden, you shall not get motivation to do it, and your efficiency comes down, and you shall be a failure. So you should develop loving your job as you love your parents, children and family.

Each work has a purpose which is a component of your success in life. Take the example of sweeping the floor. As a qualified man you might feel sweeping as a lower job and may not like to do it. Your quality of yarn depends directly on the cleanliness of the work area. If the work area is bad, the imperfections in the yarn will be more, the defects in fabric will be more, efficiency shall be low, yarn breakages shall be high, and ultimately productivity shall be low. If sweeping and keeping your work area clean is helping so much in achieving your target of productivity, quality and total production, why should you feel that work as of a lower nature. When you are cleaning yourself after answering the nature call, why should you feel bad for cleaning the work area which is just having dust.

Someone is ready to do the job which you feel like a lower job, for example, cleaning the lavatory, sweeping the floors, cleaning the plates after taking food and so on. If that job is not fit for human being like you why are you employing another human to do it? If you understand the importance of that work, how it is helping you to achieve your targets and in your progress, you shall start recognising those jobs and do it wilfully.

We love our parents and family as they are looking after us. The job we are doing is also doing the same work. If we look after our job well, the job we are doing will look after us. We are getting our livelihood by the job we are doing. If our beloved ones are sick or crippled, we do not leave them, then why should we not consider our job also in the same way and see that it is done well.

4.4.3   Enjoying work

The quality and efficiency can improve when one enjoys his work. One without interest feels the work as boring and gets tired fast although he is physically fit; whereas a person with interest does not get tired by that work. Hence one should always ask himself the following questions:

- Am I enjoying my work?
- Do I like my work?
- Which part of my work is good?
- Why it is a good work?
- Which is not a good work?
- Why it is not a good work?
- What I expect out of my work?
- What your management expects out of my work?

People may be having lot of ambitions and dreams, but should not forget that they are already in a job, and that job is providing them food and shelter. They should work sincerely and excel in their job. That can be done only when they love their job. Great philosopher of 12th century Sri Basavanna (Kannada) says "*Illisalladavarualliyoosallaru*" which means 'if one cannot survive here he cannot survive there also'. There is a proverb in Kannada "*Aalaagaballavaarasaagaballa*" which means "one who is successful as a servant can be successful as a king". We need to learn how to succeed and survive in our present circumstances rather than dreaming of a success in another world.

Once one gets married, it is the responsibility of both husband and wife to love each other, adjust themselves to the feelings and requirements of each other so that they can live happily. The same is true when we join an organization and accept certain role. We need to love our job and try to excel, by which we develop potential for excelling in any job or responsibility we take.

4.4.3.1   How to enjoy work?

Enjoying work depends on you and not on others. If you love your spouse whom you have married, you can be happy throughout the life. If you are not really loving and making a statement that you are, you cannot enjoy. You should be ready to accept some weaknesses and adjust yourself for those and make a living together. No one can be perfect in all the ways, and similarly no company will be perfect. In all companies there shall be some plus points and some minus points. It is not possible for you to correct everything. You may correct some but have to live with others. If you try you can get adjusted.

Make a self-analysis first. Try to understand what makes you happy: think about it and write it down. Take some time to make a list of the things that bring a smile to your face. When you make this list, write down everything, no matter how trivial or how irrelevant it may appear to your job. The purpose isn't to relate this to your job; the purpose is to make a list that's all about you. Ask "Why?" for each item and you need to figure out why those things make you happy. Try to find the exact thing that makes you happy. Is it actually doing it, or is it the surroundings? Repeat this for each item – dig deep. Keep asking "why" until you come up with the root of why this makes you happy. List the reasons why it makes you happy. After understanding what makes you happy, try to understand what makes you unhappy, and ask why every time. After understanding what makes you happy and what makes unhappy try to find out what motivates you. People like doing the things that motivate them. So you need to make a list of the things that make you feel motivated. Some people are motivated by helping others whereas some are by accomplishment and some by intellectual stimulation. This is not a particularly easy task but it's something you need to know about yourself. It's one of those "meaning of life" activities that will vary from person to person. What is it for you?

Examine your job and identify positive aspects. Put yourself in a relaxed state of mind and consider everything about, and related to, your job those are not negative. Now try to identify why you consider those aspects as positive and others as negative. Now try to match your positive influences. Look closely at the list of things that make you happy and the list of things you like and the things you dislike about your job. Interestingly there may be things about your job that you dislike that actually match up with some of the things you listed that make you happy. Take everything you listed about your job from both lists (like and dislike) and write those things (the ones that apply) next to items on the list of what makes you happy outside of work. On your match lists, you are very likely to have things about your job you like that match things that make you unhappy, and conversely, things about your job you dislike that match things that make you happy. Make a contradiction list that contains all those. You will concentrate heavily on those in just a bit. Look for conformations.

Just like the previous step, you will have things that match the way you thought they would. Some of the "bad" things about your job are in your unhappy list and vice versa. Make a confirmation list that contains all those. To be on the safe side, do these exercises a couple of times. Don't just hammer through it once and be done with it. This is a very, very important tool and you will use this later – if you really want to be proactive with your life, you'll use this many times in the future. Once you are completely satisfied that all your lists are accurate and complete take a pause.

Give yourself some time to let these things sink in. You'll know when you're ready for the next phase, but don't rush into it. Your brain needs a bit of time to process the information you've just developed. If you've done the steps properly, you will have some new and potentially surprising information. After a couple of days, you'll be ready.

The goal you've set is to enjoy your job. To do that, you have to be determined that you're going to make a positive psychological change. Do not believe that just because you've done the pre-work things will magically change for you. This will require a constant examination on your part of your attitudes and behaviours. Now focus on positive things. For example, if your boss is "hovering" remember that you enjoy being around people. When your phone is constantly ringing, remember that you love talking to people. When you're constantly being asked to do extra things, remember that helping people makes you happy. The objective here is to look for the things about your job that match the things you listed as making you happy and focus on those. Each time a job-related event occurs that is on your "happy" list, make a mental note "this was good because........" write the reason what you feel as appropriate.

There must be something about your job that matches your motivation list. Find those things. This is one of the things your supervisor should also know. Try to have a conversation with your boss about the things that stimulate you and see how you can get more of those assignments. Don't assume it will happen all at once; your current assignments are there for a reason and it will take time to adjust the workload, but most managers want their people to be productive and happy – that reduces turnover and makes them look better because the team performs better. When you do this, focus the conversation on the positive areas rather than the negative areas.

You sometimes find yourself thinking on how you don't like your job or you don't like parts of it. Your mind will feed on that and it grows until it engulfs you. That's what's referred to as "bad rap" and nothing good will come of it. When you find yourself doing that, yank yourself back to the positive. Think about some job-related activity that matches your happy list instead. This is not easy. Bad rap is a habit and it's hard to break, but it's so critical that you stop that in its tracks. Bad rap increases stress, can lead to depression and definitely reduces performance which will lead to poor performance reviews which will increase the bad rap... well, you see the vicious cycle.

When you have a 15-minute break or a 30-minute lunch, use that time to do the things that you enjoy doing. This is not the time to gather around with co-workers and gripe. If that's part of the culture, you need to distance yourself from it – it's hurting you. Maybe find someone that likes to walk and walk with them; talk about pleasant things. Ensure that during lunch time you associate with positive people instead of negative people. People fuel each other with their attitudes. Your objective is to increase the positive and decrease the negative. When you constantly hear people expressing negative views, your negative views will increase. Likewise, when you're with positive people, your positive views will increase. Positive thoughts are habit forming just like negative thoughts is. It works because of the way our brains are wired. It may seem useless at first and you will not see instant results – this is not a magic wand kind of thing – but keep at it. Do it every day. Post your happy list job matches on the mirror and read them every morning.

If you've done the steps and you're taking the actions, you will see results. They will be baby steps at first so you'll have to look for them. They will be there and it's up to you to find them. When you see results rejoice as you are succeeding. Every success is a positive feedback to your brain. They will start coming faster and faster. One thing though, there will always be events you don't like. Nobody loves every second of every day of their life. That's just not normal. When those happen, deal with them decisively and quickly, and then move on. Do not dwell on the negatives. Focus on the positives to come.

Many times the part you hate about your job is another employee or several of them that make life at work miserable. The thing to remember about these people is 'do not take things that are said personal'. There are some who are insulting, jealous, gossipy, and even those who may be out to get your job. Sometimes finding something you like about that person and commenting about it to them can turn someone completely around. Be interested in what

a job. Attitude should be an essential element of every job description in an organization to reinforce positives.

Organizational culture sensing should be done as a regular exercise for assessment of the organization culture and relative work values. Haphazard change without knowing the depth of existing management systems and its root causes can have severe impact on the organizational operation and also undermine strengths of a business entity.

6.2   Need to arrest employee attrition

Any organization, if has to grow or at least survive, depends on how consistently it serves the customers by providing quality goods and services. One needs to understand the changing requirements of the customers and convert them into technical parameters, design and develop the products and services and produce them at economical methods and serve the customers. This is possible when the people working are united, understanding each other, cooperating and jointly putting whole-hearted efforts for achieving the company objectives. The biggest challenge in front of the textile and garment industry is retaining the employees and developing them as one motivated team to work for the fulfilment of the company objectives. It should be noted that customer satisfaction can never be achieved without employee satisfaction. Therefore the managements must work for understanding the genuine requirements of the employees and address them. The attrition levels in Bangalore garment industry are very high; figures of 100% to 300% are very common. It means the workers just stay for 4 months to 12 months in number of cases. Similar is the case with textile industry in south Gujarat, Silvasa, Tarapore and other north Indian cities. There are very few factories in which some of the employees are working even for over 10 years, but they are exceptions. Where the staff and workers are staying for more time, the systems can be established and implemented, whereas in other cases it is very difficult.

Although the problem of high attrition is there in both garment industries and textile industries, the reasons are different. In garment industries more than 80 % are female employees, and they have number of personal problems like their family, children, health, etc. The ladies have to complete their household works and then come for their factory, and again after going home they need to do all the household works. If someone is sick at home this lady has to stay back. Again the marriage, pregnancy, child delivery and transfer of husband to a different place etc. also contribute for high attrition rates. The number of working hours is eight and normally garment factories work only one shift. In textile industry, workforce consists of mainly either migrated workers from different states or workers hailing from villages. Migrating workers lead a bachelor's life, work for 12 hours a day but leave the company when they feel tired. The workers from villages have to concentrate on their fields, and whenever there is no work at fields they come to mills for work. The problem of attrition is less in centres like Coimbatore as workers are depending on the mills, and industry systems are well-established over a period of time.

Few of the managements have a feeling that if employees are allowed to stay long, they might demand certain privileges or removing them shall be a problem later. They are adopting some methodology by which the employees themselves leave the organization. Some are engaging contractors who bring the workers and pay them, whereas the company pays to the contractor. Because of this attitude, those companies are not growing. The management is always blaming some external factors for their inefficiency.

There are instances that the management is dependent on managers and supervisors for bringing workers. The payment of the staff depends on the number of workers he/she is capable of bringing and not on the technical ability or the quality of the work done. Such companies can never improve technically. Another danger is that when this manager or supervisor leaves, all his followers also leave the company creating a big void.

As the cities grow, the cost of living becomes very high, and the employees are finding it difficult to get a decent accommodation near the work area. Their time is wasted in commuting. The commuting in buses or trains in heavy traffic reduces the efficiency as the people are fully tired by the time they reach the factory. Forget the garment industries providing decent accommodation, but are not even helping the employees to get accommodation.

6.3   Understand the employee

If an organization has to sustain the pressure of the market and economy, it can only be by whole-hearted team work and not by any other means. One can invest huge money in technology; but for managing needs, dedicated, skilled, matured and knowledgeable employees are required. If employees have to develop the above basic requirements, we need to work for them, understand their needs and treat them as our partners and not as slaves. The normal needs are wages to support decent living and to take care of the future needs of family, security, social status, dignity, growth and peaceful work environment. Employee's first and foremost needs are of food, rest, and protection from the undesired events/elements. Until these needs are satisfied, employees cannot enthusiastically concentrate on work, learning and applying new techniques, expressing self, competitive idea generation etc. Companies should make system to monitor their members in order to make sure that their basic physical needs are being met with existing compensation packages.

Few of the managements think that by paying higher wages, they can grab any one, and hence, they do not bother to fulfil other requirements. They shall be forcing the people to do a number of activities which might be against the will and consensus of the employees. Then they will not do the work by heart, and leave the company the moment they get an opportunity. They shall be working till they get another job. No sooner they get another job they leave. Therefore the treatment given to employees has become more important. This normally happens in the mills where ethical systems are not practiced, the regulatory and safety requirements are not met and humans are not respected.

The security is another factor valued more by the employees. No one is interested in changing the job just for the sake of money. When they feel insecure because of the working styles of management or bosses or feel that the job does not guarantee the required security they prefer to change.

People want status in the society. If the company in which they are working has a bad reputation of not following ethical practices, they do not like to work in that company. If the company earns a name as a friend of society, the employees shall be proud to say that they are a part of it.

People have some of their own dreams of the type of job they do. The job they do should give them satisfaction and a pride of doing it. They do not want to work like a machine without any thinking or creativity. Unfortunately the manufacturing activities in garment industries are just repetitive jobs.

Everyone wants to grow. The owner of the industry wants to grow and make the factory bigger. Similarly the employees also want to grow and hold higher positions. If the company is not helping them to grow, they shall not be interested in continuing.

The peaceful work environment is one of the utmost important factors which the employees want. No one is interested in daily struggles with various tensions. In garment industry because of the mistake of top people or marketing people for committing the deliveries without understanding the facility and the capacities they have, the people working are pressed for deliveries. The company may be ready to pay overtime wages, but the employees are not getting time to attend their family works and finally decide to leave the job and select some job where they get some time to take breath. While requirements for a high quality of work life vary from person to person, certain factors are generally required for anyone to have a high quality of work life. These minimum factors are equivalent to health, food and shelter for standard quality of life; however they are more specific to careers or jobs. For example, to have a high quality of work life, generally a person must be respected at work. Co-workers and senior-level employees must treat them fairly and politely. The work must not cause the employee any physical discomfort or mental anguish. The employee must feel as though he is doing something enjoyable or at least not unpleasant. The worker must feel the salary he is paid is sufficient for the work he is doing. Finally, the employee must feel valued or appreciated as though he is doing something of importance for the company.

6.4   What the management should do?

The most important thing the management should do is to keep some time reserved for thinking and planning the activities which is missing in number of companies. One who is calm can find solutions to a number of problems. Just by insisting to go on working cannot solve the problems, as the people cannot think and find solutions to the problem.

The management should take active part in understanding the problems and work with the people to solve it. When people know that the management is with them in solving the problems, they get moral support, and shall involve more in the activities and come out with innovative ideas to come out of the crisis.

The management should understand that the employees are their strength. Good employees can get best results out of the facility they have. Therefore, the management should work for developing the standard of their workforce, not only in knowledge and skill but also economically. They work to relieve the tensions the employees have. Good companies have systems for providing the health care facilities to family members, education facilities to children, helping employees to get houses near the factory but in a decent locality. If the tension of commuting is removed, the efficiency of the employees shall increase substantially. There are some companies which try to bring the families together so that a friendly atmosphere is created among the staff and employees. But such companies are very limited in number.

6.5   Benchmarking activities

The benchmarking systems in keeping the employees happy include providing facilities such as housing, transport, time and encouragement for the employees to enrich their knowledge either by providing training or by providing leave or adjust the work timings so that they can attend the classes, helping the employees' children to get entry in good schools and colleges by reserving some seats, helping the employees' children in their marriages by depositing some amount in the name of the children, providing complete medical facility in reputed hospitals for the family members, encouraging the innovative ideas by implementing them in the company and making it public, encouraging the employees to visit the customer's premises to understand the nature of work and the expectations of customers, inviting employees to discuss the problems with customers, providing an opportunity for the employees to visit the competitors facilities, helping the employees to form associations for exchanging their experiences and expressing their views, the senior managers visiting the houses of employees in routine to understand the problems they are facing and also to enquire the well-being of the family members, wishing the employees and their family members on their birth days and marriage anniversaries, recognizing the achievements made not only by the employees but their family members in any field and bringing it to the knowledge of all, arranging various cultural and sports activities that can help in building team spirit, encouraging the staff and workers for engaging in social activities helping the community around to develop, encouraging the staff to present papers in various seminars and conferences, encouraging the managerial staff to visit educational institutes and give guest lectures which not only makes it compulsory for the managers to read and get them updated with the developments taking place but also develops an interest among the students about the company and make them feel like joining the company, etc. Some companies have made it a policy to employ people through employees' reference.

These activities not only make the employees involved in the company activities with willingness, but also make others to prefer this company as their dream company for employment. The parents shall insist their children to join such company and they resist if anyone thinks of leaving the job. There are companies around us where these systems are followed and people are working for generations together. They are always ready to shoulder to the company in the recession by depositing the bonus amount in the company opting for taking 20% less salary for a year and give it as loan to the company, investing their money in the shares of the company and so on. When such examples are there around us, why the textile and garment industry management should not take the benefit?

The industry owners should avoid dragging staff and workers from other factories. They should avoid recruiting people for higher posts but develop the insiders and promote them. The recruitment should be only for lower jobs. When an employee fails or makes a mistake, identify the root cause. If the employee has involved wilfully in damaging the interest in the company, action may be taken. In a number of cases the mistakes happen because of the circumstances and the lack of knowledge, support and confidence. It is the duty of top management to build confidence among the employees and give them an opportunity to correct their mistakes. The systems should be transparent and impartial. We should always remember the findings of Dr. Juran that 80% of the problems are management initiated and workers are responsible for only 20%. It may be hard to digest for the people in top, but it is a fact. They should stop blaming the employees and start working on correcting themselves.

A number of companies boast of having SA 8000 certificate and claim that they are employee friendly. If you study in detail they are all the basic requirements as per our factory Act and other related Acts, and all are supposed to follow it. As our industry owners are not sincere enough to follow the law of the land which is providing food, shelter and helping them to survive and grow and as our government officials are corrupt and not implementing the law and regulations, the foreigners are not having faith on our people. So they are insisting on SA 8000. The management who are not abiding the rules and regulations must be put behind bars but our government is not strong enough to do it, and our people are not educated enough to demand it.

6.6   Some thoughts for the companies with migrating workers

At companies where majority of the workmen and junior staff are out siders they normally are living a bachelors' life. They need to commute for attending the work as the companies are not able to provide accommodation near the factory because the factories are located at industrial estates away from the city. They get tired especially while having 12 hours work. The following are some suggestions that can be considered for such situations:

Providing housing facilities near the factory either by constructing them by company or by taking them on lease.

1. Providing loan facilities to employees for purchasing a home in nearby areas.
2. Providing loan facilities to employees for purchasing a home in nearby areas.
3. Providing dormitory to bachelors.
4. Providing education loan for employees to enrich their qualifications.
5. Encouraging employees to bring their families.
6. Interacting with good schools to provide seats for the children of employees so that they can comfortably think of bringing their family.
7. Providing jobs to the spouses or helping them to get a job by the influence of the management.
8. Providing loans for purchasing vehicles and other home appliances at nominal rate of interest.

9. Providing special loans/monetary help for those getting married and willing to bring their family.

10. Strictly following the safety regulations and other legal and regulatory requirements to ensure security to the people working at the company.

11. Developing the systems as per the concepts of total quality management and ISO 9000, which boost the morale of the people and build team spirit among the employees.

12. Involve employees in small improvement projects and recognize their efforts in bringing improvement. This encourages them and make them feel continue here.

13. All management staff (supervisors and above) are to be trained for enhancing their management skills, recognizing and respecting people down the line, mixing in the team and leading from within (concepts of thread in the garland).

14. Encouraging the staff for becoming members and taking active parts in professional bodies like The Textile Association, Institute of Engineers, Asian new work for quality, and Indian society for quality and so on. This makes them more knowledgeable and also interested in doing their jobs willingly.

15. Sending people for training outside the factory and using them for training other employees on various matters of interest for the company.

16. Identify the people interested in sports and build teams. Encourage them to improve their skills in the games they are interested by sponsoring matches and also sending them for open competitions.

17. Keep the working area always clean and stabilize some product as regular working products. Good working conditions and stability in working develop interest in people to come for work.

18. Considering development of an employee cooperative society for consumable good and a co-operative credit society. This brings involvement in people and they like to continue in the company.

19. Promoting people from within, and recruiting only fresh candidates and training them for jobs.

20. Having get together of family members of the employees once in a year and highlighting the company activities.

21. Providing a facility similar to LTC for the family members of the workmen to visit once in a year or once in two years.

22. Flexible work time especially for administrative jobs and job rotations for some of the monotonous jobs can be considered.

7

Key result areas and performance indicators of work quality

When we claim that our work quality is good, we should be able to demonstrate it by some results. Following are some examples which can be used for demonstrating the level of achievement of work quality in textile mills. You can add more to this list.

1. *No loss of production or delay in activities due to shortage of human resource.* If the work life quality is good, people come for work with interest and there cannot be any shortage of workers.

2. *Reduction or no complaints relating to noncompliance of regulatory norms.* Respecting the people, the community and the law of the soil is imperative for achieving good work quality. People work whole-heartedly where their community and their country are secured and respected.

3. *Reduction in legal expenses.* If one is respecting the law and work as per that, there need not be any expenses relating to legal aspects. One who tries to violate the law to get short-term benefits has to spend for lawyers and court.

4. *Reduction in absenteeism.* Reduction in absenteeism is an indicator that workers and staff are happy with the work atmosphere in the company.

5. *Reduction in strikes and disputes.* When the work quality is good, there shall not be any disputes that can lead to a strike. People shall discuss with their seniors and colleagues and agree upon the actions to be taken and work as one team.

6. *Reduction in employee turnover.* If work atmosphere is good, people would not like to leave the organization unless for their personal reasons.

7. *Reduction in employee expenses as a percent of sales turnover.* When the work atmosphere is good and people are working, the productivity and quality shall improve and the sales realization also shall improve. Number of expenses relating to recruitments, appraisal and supervision can be cut down. The ratio of employee expenses to

total sales turnover shall reduce. It can be seen in good companies where the employees are respected and treated as partners, the salary or wages are less compared to the mills where employees are not treated well. They are offering high wages and salaries to their employees to attract them and to retain, but are not successful.

8. *Availability of complete information in the personal records regarding the employees.* It indicates the interest that the HRD personnel are having regarding updating the information regarding their employees.

9. *The time taken for recruitment.* If the work quality is good, there shall be number of people eagerly waiting to get a job in that company. Hence, in the event of a vacancy arising, it shall be filled fast.

10. *Expenses for recruitment as percent of sales turnover.* When people are not leaving the company, the recruitments also shall be low. Certainly the expenses for recruitment shall be negligible.

11. *Reduction in number of grievances.* When people are bonded and are working as teams there shall be no grievances.

12. *Facilities of reading room and library utilised by the staff.* Where the employees are enthusiastic, they strive to come up. They spend some time for reading news related to the industry and the technology. The present problem in textile and garment industry is that the staffs, both technical and administrative, do not spend any time to learn by themselves by reading books, technical or commercial magazines. Hence they are not getting the information regarding the changes and developments being taken place.

7.1   Training

1. *Reduction in market complaints relating to worker controllable errors.* A good work culture allows people to concentrate on their works, and people can be trained well. This reduces worker controllable errors.

2. *Improvement in worker efficiency.* No need to explain this. When people are working with heart, the efficiency is bound to increase.

3. *Reduction in wastes relating to worker errors.* When work is done with interest and concentration, there cannot be any wastes that are avoidable. Only process wastes that are set shall be there.

4. Clarity about the job content, the responsibility and authorities among the employees reduces confusions and people get empowerment. The jobs shall be done without any obstructions.

5. *Clarity about the product requirement among the employees.* When people are clear about the product requirements, they strive to achieve that.

6. *No loss of production due to shortage of required skills.* People voluntarily come forward to learn and enhance their skills where the work quality is good. A trainer has to give only guidelines and the rest shall be done without any efforts.
7. *Improvement in average skill level among employees.* When people are enthusiastic, they learn and acquire knowledge and skill, resulting increase in average skill level.

7.2   Stores

1. *Reduction in complaints relating to non-traceability of materials in time.* The work quality of stores is measured by the quickness in which the materials are issued when a requisition is given. Keeping the materials in their specified location and updating the stock readings as and when the materials are added and materials are removed can be done only when the people working in stores work with dedication and responsibility.
2. Time taken for preparing the goods receipt notes (GRN) and getting the materials approved is an indication of work quality of stores.
3. *Zero difference in stock shown and the actual stock in stores.* Normal problem seen in majority of stores is the difference between actual stock and book stock. Entering the figures in system and in bin card as and when materials are received and issued can be done only when the people working are dedicated.
4. *Reduction in wastes and damages due to handling/storing at stores.* Store is meant for keeping the materials safely. If the materials are getting damaged in stores, it means a poor work quality.
5. Clarity about the shelf-life of materials, storing them accordingly and issuing by referring to expiry date is one of the main responsibility of a store keeper, which reflects the quality of work at stores.
6. Clarity about the materials requirement for different sections of the organization and arranging the materials as per section is an indicator of quality of work at stores.
7. *No loss of production due to shortage of required materials in stores.* Store is supposed to issue the required materials in time so that the production does not suffer. Bringing extra materials from stores and keeping it in production area is a poor work culture.

7.3   Accounts

1. Speed of giving the accurate information relating to monetary position is a very important requirement of the management for taking decisions.
2. Time taken for preparing the accounts statements is an indicator of work quality at accounts.
3. Zero difference in accounts shown and the actual money in hand is the primary requirement for the efficient working.
4. Wastes and losses due to improper accounting and non-fulfilment of statutory and legal requirements are indicators of poor work quality.
5. Clarity about the accounts and appropriate allocations of funds for various activities reflect on the work culture at accounts.
6. Clarity about the money requirement for different activities of the organization.
7. No loss to the company due to mistakes or delays in accounts sections.

7.4   Production

1. *Production achieved in time.* If the work quality is good, the planning and coordination shall be good; and hence, people shall ensure that work is done in time. They shall monitor the process stock and balance the back process machinery. A continuous interaction in supply chain can be seen which ensures that customer gets materials in time.
2. *Quality levels of the products.* When people are working with cool mind, chances of mistakes are less. When mistakes are not there, the quality is bound to come.
3. *Quality of packing.* Packing is meant not only for safe guarding the materials but also for presenting it to customer. A good packing not only should focus on safety of materials but also should be attractive. The customer should be provoked to purchase your material. If packing quality is poor, i.e. the dimensions are not proper, the markings are not in line or not attractive, the colour of packing materials and the design on the package are not attractive, the straps are loose, the packing is not tight, then the customer would not like to see the material inside. Therefore the quality of packing is a good indicator of work quality.
4. *Reduction in rejections.* The rejections may be due to any reason like human errors, improper communication from top, improper

maintenance practices, improper selection of raw materials, improper setting of the machines, improper maintenance of temperature and humidity, and so on. If the work quality is good, the rejections shall be less.

5. *Increase in production efficiency.* With good work quality, people working are less stressed and can organize their works efficiently leading to higher efficiency.

6. Reduction in wastes of all type like process wastes, product wastes, under utilization of men, under utilization of machines, power wasted, humidification plant uncontrolled, compressed air wasted, power wasted, water wasted, excess inventory, over designing and so on. When people are not stressed and have cool mind, they can think on the activities that are not adding value and can eliminate them.

7. *Increase in product realization.* Higher product realization is achieved when wastes are reduced and, wherever practicable, are reused without affecting the quality. This requires a calm approach. Only by forcing people, it is not possible to achieve quality with higher product realization.

8. *Reduction in cost of manufacturing per-unit production.* Cost of manufacturing can be reduced when we can apply our mind and identify wasteful expenses. Therefore reduction in cost without compromising quality and delivery in time is a good indicator of work quality.

9. *Accident-free performance.* Accident takes place when people are not alert. Over burdening with various thoughts is one of the main reasons for losing concentration while working, which may lead to accidents.

10. *Reduction in stocks of finished materials.* A good planning and coordination with marketing and customers can reduce the stock of finished materials. Treating customers as partners, understanding their real needs and focussing all activities to meet customer needs can result in lower stocks of materials produced. Just to show higher utilization or higher production if we go on producing the materials without understanding the markets, we shall push the industry toward recession.

11. Mix up of materials, poor quality due to fluff accumulation, stains in yarns and fabrics are indicators of poor work culture.

12. A failure in tracing the materials and non-reconciliation of materials is an indicator of poor work quality.

13. Keeping excess stock of work in process and spares is an indicator of poor work quality.

14. No damages to materials, floor, walls and for human beings by the material handling practices is a good indicator of work quality.

7.5   Marketing

1. Reduction in marketing expenses as a percent of turnover. If quality of work is good, quality of product and services also shall be good. Marketing people need not to make any effort to search customers, but customers themselves shall come for purchasing.
2. *Customer retention.* Customers are normally not interested in changing their suppliers. They are forced to change the suppliers because of poor work quality of marketing personnel. Hence, retention of customers indicates good quality of work at marketing.
3. *Reduction in market complaints due to non-product issues.* It is a known fact that only 4–6% of the market complaints are related directly to product quality, whereas it is other factors that create dissatisfaction among customers. A good work quality in marketing can reduce these dissatisfactions.
4. Quality of display in retail showroom is very important indicator of work quality as it depends mainly on the interest and involvement of sales person and not on any technology.

7.6   Maintenance

1. Reduction in expenses for maintenance while maintaining zero breakdown and product quality is a good indicator of quality of maintenance.
2. Reduction in power consumption, reduction in consumption of compressed air, and reduction in steam consumption are all indicators of a good quality of maintenance.
3. Reduction in fires, breakdowns, consumption of diesel in power generation, steam loss in transmission, power loss in transmission, air leakage, etc., are all indicators of work quality in maintenance.
4. Increase in mean time between breakdowns of each category is a good indicator of work quality of maintenance.

7.7   Purchases

1. Providing the required materials to users in time is the main responsibility of purchases while maintaining lowest possible inventory.

2. Procurement in time at lowest price, minimum stock level, and lower rejection of the purchased materials are the key indicators of work quality of a purchase section.

7.8   Design and development

1. If designers are able to understand the likings of customers, then the percent of designs approved and converted into orders shall be high.
2. New designs and styles successfully introduced and commercialised is an indicator of work quality of designing.

7.9   Production planning and control

1. Planning the production activities to deliver the materials in time to customers and at the same time having maximum possible utilisation of available production capacity is an indicator of work quality of production planning and control.

The above points are just some of the examples, and each mill has to define and work out the key performance indicators depending on the company objectives.

8
Five golden questions

The textile and garment industry is working in a highly competitive environment. To become a winner we should always be competitive and keep performing at highest levels. However may be the process being managed, we need to understand the basic purpose for which we are working, how we are going to achieve that, what results are expected and how we are likely to fare in the competitive environment. A number of quality management systems are developed, and all have the same intention of making the company/person stronger and competitive. Each system has its own policy and methodology. The promoters of the systems claim that their systems are the best. The implementers when study different systems are in a fix, as all are good, and how to integrate them is not clear. There is a need for simple concept, which can help everyone. The following five basic questions named as golden questions, if answered promptly, can help in achieving the targets. These questions are to be asked to self and not to anyone else. A sincere and honest reply is required.

8.1   Whether we have a procedure?

The first question is "whether we have a procedure?" for doing any work, there should be a procedure, i.e. a defined method of working which is established and capable of giving the result when followed. The procedure may or may not be documented, but should be in practice uniformly throughout all the time. Everyone should be working in that style without fail all the time. Sometimes we see a documented procedure in place, whereas the people are not following. The documents are kept securely in cupboards. Each one is doing the work in own way. In such a case it is "no procedure". The procedures should have been established by judicial studies and logical thinking and not just by some one's ideas. The procedure should have been tried and established before declaring and documenting it as a procedure. Then only we can say that we have a procedure. First verify whether there is an established procedure for the work you intend to do.

8.2   How do we ensure it as the best?

Only having a procedure is not adequate. It should be the best and suitable for the activity in hand. The procedure should address the objectives, and be suitable to achieve that. We need to periodically review the procedures to find their suitability considering the changing environment. Therefore, the procedures should be tailor made for the situation, and not to be copied from others, just because they are doing well or they have some reputation. There might be different ways and means to do a work or to achieve the objectives. We need to understand our culture, competency, facilities, capability, resources, time, etc., and verify the suitability of the procedure to our environment. We need to have a procedure best suited for our environment. There should be a system of periodic review of procedures to ensure their continued suitability considering the changing environment.

8.3   How we are implementing it?

Writing a procedure is very easy comparing to implementation. It is more complicated when it has to be implemented organization wide. When we say implementation, it always means organisation wide implementation without any excuse. In Indian culture maximum respect was given to those who practiced what they preached, and were called acharyas (teachers). It requires willingness and determination to do the work, education and training to all so that they can understand and do the work. People should have confidence and belief in the system to follow it without hesitation and interruption. The main reason for the failures in system is the disbelief among the implementers starting from the top. A whole- hearted implementation demands hard work and discipline. There is no substitute for it. The implementation means doing the activity as per the agreed procedure and reviewing them on-line and correcting as and when a deviation occurs.

8.4   Did we get the results as anticipated?

Any procedure as stated earlier should address the objectives. If the procedure is evaluated and found as the best to achieve the objectives, and if we are able to implement it in real sense, then we should get the results that were anticipated. If the results are not obtained, it means that either our procedure is not suitable or we have not implemented as required. Hence, monitoring and measurement is very important. We need to learn measure everything, and then only we shall be able to say whether we got the result as anticipated. If we do not know how to measure, then we will not be able to monitor it.

they have to say. Ask questions about them or their life. Many times that person will begin to not know why they like you and they will change the way they react to you. When you have found someone who is a constant problem, this is evident to everyone else too. Don't get caught up in their ways.

4.4.3.2   *Whether the other job is good?*

You may do all the exercises explained above and feel it will not work. If these don't work; if you are 100% sure beyond the shadow of a doubt that you will never, ever like the job you're in, and you're going to find a new one, use what you learned doing these steps to evaluate the potential new jobs. Make sure your next job fits what you like. Realize that your attitude impacts others as well as your job performance. If you do not have an attitude of understanding and adjusting yourself to a job, you may find the new job also as boring as your old job. Remember that the grass is not always greener on the other side. Sometimes you can change a job only to find that the new job has an employer who is very difficult to deal with or worse employees, even to have double the work load or less pay or benefits. Make your homework before you make that change. Get to know more about the potential new job and the people you will be working with.

To achieve a high quality of work life, it is essential to choose a job that fulfils your needs. First, you must determine what those needs are. If you want a job that engages your mind and challenges you, it is important to understand that in advance so you can earn the qualifications that will allow you to obtain such a job. It is helpful if you choose a job you are interested in; you need to consider what your interests are and research jobs within those areas. Make a list of things you are looking for in a job and speak with a career counsellor or attend career fairs to determine which jobs are most likely to fulfil those needs. Finally, pay attention to your interaction with existing employees when you go for interviews – the way you are treated by your boss and co-workers will have a tremendous impact on your quality of work life. You will want to ensure the culture of the business matches your own comfort level.

Unfortunately, despite their best efforts, some people find themselves with a low quality of work life. They may be forced to take a job they don't enjoy because of personal or financial circumstances such as a lack of options or education or qualifications. For those with a low quality of work life who are unable or unwilling to change jobs, it is important to cope effectively with the situation. Unhappy employees can attempt to improve their quality of work life by choosing to focus on the positive components of their jobs. A shift in mindset to focus on the benefits, even if those benefits are minimal,

can improve the quality of work life. Unhappy employees can also explore opportunities to speak to co-workers and management to remove factors that reduce the quality of their work life, if possible depending on the job situation.
"Love your spouse by heart and be happy life long".

4.4.4   1H, 2H, 3H and 4H concepts

Earlier, for doing a work hand was considered to be the most important among all the organs of human being. Hence the terms "hands required", "production loss due to hand shortage", "number of hands engaged" etc., found place in the manufacturing industry. In the family circle, the 'earning hand' was the word used for the person earning for the family. As the technologies developed only hand work was found not sufficient, and people were asked to use their head while doing the work. But the recent revealing is that when you work with heart (dedication in work) while using your hand and head effectively, you get the best results.

    1H = Hand

    2H = Hand and head

    3H = Hand, head and heart

There is a relation between the extent of happiness to success or failure. Sometimes you might not be happy with a success, compared to a failure. It all depends on the challenge you faced. You might not be thrilled when you get a success without any challenge, compared to a failure in the last stage after overcoming a number of challenging hurdles. Losing a cricket match with one run in the last ball of a final match in the tournament gives more excitement than a clean sweep on a weak team in the first round. When you love a job, do it with dedication and interest, remove number of hurdles, you shall get job satisfaction. The happiness you got in the job in spite of its failure shall give you energy to work again with all the 3H together, and finally achieve the result.

When we analyse some of the achievers like terrorists, robbers, smugglers, cheats, etc., they also love their job and work with 3H, but we cannot accept them. What they miss is humanity or human values. If an achievement is done while human values are respected, it is real achievement. We need to add the 4th H, i.e. the human values for all our activities and then our work will become immortal.

Producing the goods and services to fulfil the customers' needs and expectations is normally insisted to make the company competitive. Whereas one should understand that respecting human values and ethics are also very important for getting the support from the society for the sustenance of our business. We are a part of society consisting of human beings and we need to respect the human values. All our employees, suppliers, customers and other stake holders are all human beings. So adopt **4H** to become successful all the time, i.e. **Hand, Head, Heart and Human Values**.

4.4.5   Three steps to enjoying work

Many people who work in corporate office environments tend toward high levels of stress and anxiety, while others happily enjoy their work life. How can the same environment with the same stress attributors cause such different responses? The obvious answer is that people who enjoy their work do things differently from those who stress out. But what do they do differently, and how? Richard Lindesay identifies three patterns with stress in corporate environment which is explained as follows.

State: People have a base level of stress that they can comfortably deal with. If their day-to-day stress is at or below this level, they get by fine. But if things get too stressful and above the level they can comfortably deal with problems occurred.

Those with a low base level of stress can generally deal with more than those who have a high base level of stress. Those who have a high base level of stress only need a few things to happen before they're tipped over the edge.

As people go over their comfortable level of stress, they deal with it in different ways. Some people will do something to distract themselves, such as smoking, eating, drinking and going for a walk. Those who enjoy their work notice that their stress is getting too high and reduce it without the need for the distraction.

Method: Those who enjoy corporate work and have low stress levels tend to have systems to managing their workload. Such systems help them plan to avoid stress by having ways of dealing with incoming information, planning tasks, and managing information overflow.

In corporate environments there are many different ways that information comes to people. Information comes from meetings, through email, phone calls, text messages, and people dropping by the desk. Those who have low stress tend to have ways of gathering this information and dealing with it appropriately. Now with all of this information there is invariably going to be work to do. Effective time management methods go a long way toward reducing stress and anxiety, as people then know what is important now and what can be left until later.

There is often a lot of information that needs to be kept and stored for later use. Those who have low stress levels tend to have ways of organising this information in such a way that it can be easily picked up later and understood without too much effort.

Attitude: People who enjoy corporate life tend to have flexible attitudes. One of their main attitudes tend to be one of being light-hearted and easy going, and therefore they don't get too stuck in one way or another and can instead adapt as necessary. Such light-hearted people also enjoy themselves more, and therefore those around them tend to also enjoy themselves in their presence. If people are able to see enjoy themselves and see humour in things that would usually bother them, they tend to not bother them as much.

Richard Lindesay's approaches focus on reducing the base stress level, therefore making it so that the person can deal with more stressful things before it becomes a problem. Then give the client ways of quickly dealing with stress when it gets too high, without the need for distractions. He often gives guidance on input, task, and information management. There are various methods used over the years which are adaptable to all sorts of different personalities and styles. He suggests working with clients to find something that suits them.   Finally having an air of humour and fun around the work and encouraging those work with to find humour in their difficulties, helps them think of them differently and come up with better solutions.

5

Quality people: A key to excellence

The globalization and opening up of world markets along with rapid technological developments are the salient features of the present economy. Severe competition between industries and within industries can be seen world over. Everyone is trying to be competitive. Installing the latest technology and introducing new products with special features are the normal strategies adopted. Huge amounts are spent in installing the laboratories, plant and machinery with state-of-art technology, and highly qualified managers and assistant managers are employed to run the plant. Efforts are made to get themselves accredited to some internationally accepted quality management systems in order to attract the customers. Huge amounts are spent for advertising the products and the brand. They try their best to project themselves as "quality people". The image of "quality people" and "quality company" are felt as very essential to attract the customers to be in the market and run the business successfully. The textile and garment industries are not an exemption.

5.1   What do we mean by quality people?

Everyone dreams to excel in his/her field and does some efforts to make the dream a reality. One who succeeds is branded as a 'quality person' and is benchmarked. Similarly all organizations also like to be successful in their ventures and do a lot to get the reputation of a 'quality company'. The efforts include improving the conditions by adopting latest technologies, cutting costs wherever possible, recruiting people with high qualification and experience, offering good remunerations and packages to attract and retain competent staff, training and educating people, benchmarking and adopting systems that brought success to other leading companies, trying to develop a culture compatible to the requirements and impressing the customers and public by various means. However, one can always see a gap between the achievement and the ultimate expectation. The expectations go on changing as we achieve and there is no limit. People get frustrated when the expected results are not achieved. We often hear the reasons for failure, like 'our people are not good, they are lazy, wasting time and finding an excuse for all, searching for an opportunity to relax, do not have any attachment towards the company, unprofessional, illiterate, do not adopt to new technologies, not at all proactive', etc. On the contrary the people working in the organization say "mismanagement, they don't respect our words but do whatever they feel as correct, no respect for human beings, no planning while spending money, interested in building their homes and not the company, and so on".

People like to project themselves as correct, efficient, knowledgeable, sincere, ethical, hardworking, result oriented and do a lot to impress others. Everyone wants a reputation for self. The companies also would like to impress the customers and the public in order to have continued business by projecting themselves as 'producers of quality goods', 'providers of quality services', 'partners in the uplifting of community', 'guardians of human values', and so on. They advertise in different media, prepare brochures with the photos of latest machines and equipments they have, display a copy of ISO 9000/ISO 14000/SA 8000 or any similar certificate to claim that they are quality and ethical people. They also quote names of some international brands for which they have supplied or got an enquiry in order to claim themselves as quality people and impress the customers.

People feel proud while working with latest machines and equipments. Customers feel that a company with latest technology and best infrastructure is capable of giving consistent good quality. The customers expect a consistently high-level quality product at low price, delivered at shortest notice because of the latest state-of-art technology adopted. The management also ensure best possible raw materials and accessories are purchased (even by paying premium prices), and used so as to achieve the best quality products. Inter-company comparisons are made for quality, and efforts are made to always remain on the top. But we see that in a number of cases, the companies are struggling to survive. The members of the management, who have invested huge amounts, are grumbling that even after investing huge amounts on new machines, employing highly qualified professionals by paying very high pay package, spending huge amounts for getting various certificates like ISO 9000, ISO 14000, SA 8000 etc., there are no returns. The market did not pay any premium for the quality goods made with latest machines; it goes by the lowest quotation and whatever earned is just sucked by the banks as interest. Workers demand higher wages to work on latest machines thus washing out the benefits of rationalization. The so-called state-of-art technology really did not help the investors in number of cases and we see closing down of business because of losses. Then what is the use? What is the meaning of quality people? Why we are not getting the success we wanted?

5.2   Musician and a professional

Once I had been to a music concert by a well-known *veena vidwan* (artist). The music was melodious and I could not feel the time running off. I could not believe myself that I had spent three hours there. The way in which the *swaras* were coming was really astonishing. As a textile engineer I was just observing the speed in which the fingers were moving and how they were pressing different strings. There was no confusion of any type. The *vidwan* was never seeing the strings or the reeds and his eyes were almost closed. I wonder how he was able to achieve this perfection. I cannot produce even a single *swara* properly from the same *veena*. I was thinking on the same all the night. The music of that *veena* had filled my ears. I was able to recollect the melody of the concert and enjoy the music again and again.

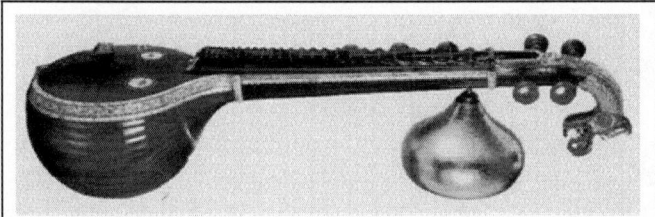

Veena – the musical instrument

What is a *veena*? For an engineer like me, *veena* is just an instrument made by a carpenter. *Veena* is one of the oldest instruments known designed with some calculations to get different sounds from different places. People in India know *veena* from time immemorial. Goddess *Saraswati* is pictured as always holding *veena* in her hands. All *veenas* follow the same specifications for strings and reed calculations so that anyone can play. It is considered as a complete instrument providing the basic components: *shruti*, laya and *sahitya*. Its main attraction is the mellow tonal quality capable of evoking a meditative atmosphere. If it is the case, why there should be so much of a difference in quality between musicians? Is it not possible to get the same quality music by all? Who should be blamed, the artist, the guru or teacher, the instrument designer or the instrument manufacturer?

5.3   Technology and human involvement

A scientist develops concepts and logics and invents new instruments. An engineer adopts the concepts given by a scientist for developing the instrument or machine to suit the requirements considering the practical aspects and safety. An engineer makes the instrument commercially viable. Although the scientist along with the engineer together develop technology that is user friendly and capable of giving a better result than the earlier adopted technology, neither of them are the people to operate it. The one who designed *veena* and the one manufactured are scientists and engineers; they are not the players. They know some basics and can explain the functions of different strings and reed. But they cannot play the *veena* and mesmerize the people like an expert musician who can take the audience to a different world. The musician can play the same melodious music irrespective of the age of the instrument as he knows its tuning. Same is the case with any industry. Just by investing huge amount on latest technology need not give the expected results unless the people know the art of handling it. It is the dedicated efforts of people who are involved in the activities having skill of handling the situation and working in harmony with others.

It is normally seen that the industry users of the machinery depend much on the manufacturers for the maintenance of their machines including the scheduling, replacement of parts, overhauling and settings, whereas it should be the work of shop floor technicians, as they are the users. For tuning a *veena*, the musician does not go to *veena* manufacturer. If shop floor technicians take interest and practice the tuning of the machines, they should be able to make it give the required quality. The fear of failure or of spoiling the costly machines make the management think of taking the help of machinery manufacturers, but by that, they are making their own people inefficient and useless. One should remember that unless he jumps into water he cannot learn swimming. One should be self-sufficient to protect self and come out successfully.

An artist while giving a concert cannot afford making a mistake. He has to practice for that. This is known as *riyaz*. Practicing daily in the correct method without making any deviations requires strong determination and a high level of concentration. The practicing is hard, but the musician does not give up. He does not get distracted by the words of others. Practice makes the man perfect and so also the musician. His music is perfect because of his practice. Consistent quality and productivity in a textile or garment industry can be achieved by continuous practice of good work systems.

So many times I have asked myself. Why we, the engineers and professionals working in an industry, cannot concentrate on our work like a musician does for music? Why we are not having that dedication? Why we are getting distracted by other activities and get carried away? Probably because we in the industry working as professionals lack in ownership and feel that we are working for others, the owner of the company. We feel that the owner shall make profit by our hard work, but gives a small portion to us. We are always working out the returns that we get because of the work we did comparing to our counterparts in the industry. We concentrate mainly on satisfying the ego of the owner or the top boss rather than on the actual work we are supposed to do. We feel secure when the boss is satisfied.

The musician is not singing for others. He sings for himself. He wants to achieve supremacy which can make him satisfied. He enjoys his music. We are working to satisfy others, either our boss or someone else above us. Our concentration is on the returns and not on what we deliver. We need to love and enjoy our works.

If we need to succeed in our assignments like a musician in music, we need to be dedicated in our works. Dedication is possible if we are getting satisfaction in our works. In our case, we are trying to satisfy someone, either a boss or a customer, who shall never be satisfied. They have lot of expectations, but never express fully what they wanted. After seeing the results they add something above it and say that was what they expected. Therefore the one who is working for such owners and customers loses interest. A musician also enhances his targets, but after confirming that he achieved what was targeted earlier and that he can sustain. He practices without seeing the time; as for him achieving the result is the prime factor. In our case time is important and the reward we get vis-à-vis the time and efforts spent on it. The musician wants to achieve his enhanced targets and it is his music. There are no rewards or loss of pay for the time he spends for riyaz. What he aims and gets is 'self-satisfaction' which is non-measurable in terms of value.

5.4   Dedication and sage

A good performance or a consistency in performance is a result of consistency in all the activities. They include understanding the requirements of customer (stakeholders), designing the product and processes to achieve the requirements, educating all concerned and working out suitable methods of practicing, motivating the concerned in getting fully involved in the activities with a feeling of belongingness and ownership, not getting distracted from the activities by external influences, a determination to complete the task successfully and whole hearted support by top management and colleagues. Dedication is the key word for success.

A sage

One should work like a sage, concentrating on one activity till he gets the result. A sage shall get up only after getting the result. Sages are not distracted by even hunger, rain, storm, wild animals, family affairs, politics or any worldly attractions. We are getting distracted for various reasons, and in a number of cases with the matters completely irrelevant to us. The lack of clarity in objectives to be achieved, improper or ineffective planning and the confusions in fixing the priorities, authorities and responsibilities are the main reasons. The priorities are always changing, and the decisions are not taken in time.

The people get frustrated and lose interest when they fail to achieve the required results or when they see their competitors overtaking them. A sage shall concentrate only on his target and never get distracted by the failures.

We really do not know what we want to achieve. We are always seeing the competitors and those making money are benchmarked. Our bosses insist us to follow what they say or feel as correct, and hence we do not get ourselves involved in the activities and are not dedicated as a musician or a sage. Without dedication excellence cannot be achieved.

We do not have the courage and confidence to correct or educate the bosses when their decisions are likely to fail in getting results. We are always afraid of the ill effects but never understand that the boss is also interested in getting results and becoming success. The sages never afraid or feel bad of telling what they felt as correct.

A Sage need not always sit in a forest and do meditation. In earlier days, sages were going to forest as they could get a calm place to think on the problem they wanted to solve or find solution. One can concentrate at his workplace itself, like scientists or engineers. The people those have developed new successful gadgets have all concentrated while developing them. They all worked like sages. If they can work like that, we also can work.

5.5   Acharya (आचार्य)

In Indian tradition, the word *acharya* is used with a great respect for the person practicing exactly the same as he claimed or preached. The word '*acharya*' means the work procedure or the work actually practiced and '*Vichara*' means the opinions or the values. One who exactly works as per his '*vichara*' is an '*acharya*'. Bhishma was called as an *acharya* although a *kshatriya* (warrior) by birth and practice. He did not get distracted by the kingdom, wealth and luxury offered to him, but remained as a bachelor and worked only for safeguarding the interests of the country for which he had taken an oath. *Acharyas* are not distracted by the luring events happening outside that can be a threat to the basic purpose.

An *acharya* is not just practicing his preaching but educating others to get benefit of his preaching. Excellence is achieved normally by *acharyas* as they never deviate from the procedure laid by them. They are very strict and never lose the heart by the obstructions in their path. They are clear as to why they have laid that procedure and how it helps in achieving the objectives. An acharya is considered as a guru, the one who shows the correct path for all. People shall be happy to follow him.

5.6   Yatha raja tathaa praja (यथा राजा तथा प्रजा)

Children, as they grow, watch their parents and other elders and try to imitate them. Similarly the people watch their leader. Both follow the footsteps in their idol's (parents or leaders) deeds and not what is preached. If I want my assistants to be honest, hardworking, sincere, proactive, knowledgeable, cooperative, etc., then I should be all those. I cannot expect my assistants to be better than me in attitudes and behaviours and still remain as my assistant. The moment they are better, they leave me out and join where their boss is better than them. Normally people do not like to work under someone who is not superior to them. So if I have to become successful and get full cooperation of my people in all my works and working in the style I explained them, I should remain superior. I should become an acharya. Money can produce number of followers to say 'Yes- Yes', but not the one to really understand and do the work which is correct and is going to bring me up.

If top man is committed others follow

We should always remember the statement made by Sage Vasista while crowning Sri Rama. He said '*Yatha Raja Tathaa Praja*' meaning that the attitudes and behaviour of people in a kingdom shall be just a reflection of their king. Sage Vasista was the guru of Sri Rama. He told Rama very clearly that "If king is good, people shall be good and if he is bad they also shall be bad. People shall follow the deeds of the king. Therefore the leader should be a role model". Rama strictly practiced the good systems designed for the wellbeing of his subjects and implemented. He was well known for his simplicity, determination and valuing the views of his citizen. He never claimed himself to be a superman or god. People considered him as a role model and referred him as *"Maryada Purushothama"* [Maryada = one who does not violate the rules or go out of his limits, Purushothama = Superior most man]. Even today, after 51 centuries people are worshiping him as god. People wish their leader to be like Rama. This holds good for any person, organization or nation and is true all the time.

The top man if does periodic reviews, without fail, of all the activities considering the Mission, Policies and Objectives, the departmental heads and sectional heads also shall do the reviews. They shall correct the deviations whenever it is found. If top man follow the legal and regulatory requirements and insists on its compliance, others also shall adhere to it. If top man is quality conscious others also shall be conscious and customers shall get quality products. If top man is selfish and trying to fool others, the people down the line also shall fool others including the top man. If we need our company to be successful we should be true to over selves.

Any quality system is top driven. If the top man is not a 'quality man' then the company cannot be a 'quality company' and the people cannot be 'quality people'.

5.7   Discipline

Discipline is very important for success. In kindergarten we had studied the story of hare and tortoise. Without discipline, the hare lost to a tortoise. The story is not only for kindergarten. The same is applicable for us also.

Without discipline the hare lost to a tortoise

Highly successful people like Sir. M. Visvesvariah, Mahatma Gandhi, J.R.D Tata, Ghanashyam Birla, Lakshman Rao Kirloskar were all known for the discipline in their life. They could achieve what they wanted and became a model to the world. They all faced lot of hurdles but never got disappointed. They were never distracted from their path. They achieved what others even could not dream.

We expect our assistants to follow the instructions given by us, but in the meantime, we ourselves violate the rule made by us or by our superiors. We not only violate but try to justify our deeds by saying that 'Rules are meant for breaking; then only developments can take place'.

In a company I saw a big board stating 'Think unthinkable', but if someone gave a different opinion than the boss, the next minute that man is out. Similarly we see big hoardings stating that "We empower people", "Our people are our strength", "Let the light of thought come from all the directions" etc., claiming that we respect our people. Some companies go to the extent of calling all their employees as "associates" instead of calling employees or workmen. If we are true and believe in these statements we should give free hand to our assistants so that they can do something. We should tolerate if some mistakes are done while implementing or taking a decision with an intention of achieving the results. But we do not give them freedom. We penalize them for their inadvertent or technical mistakes also. This indicates that all slogans are only for decoration and to impress the public. There is no need of a slogan or a display board where the top man is mixing with people and working with them by respecting their views. People who work there know their boss very well. They can demonstrate their faith in the management in their deeds that are visible. Even an outsider can understand the culture of a company just by interacting with few people either inside the company or outside.

We insist people to chant the quality policy daily, but never insist on implementing as suggested in the quality policy. If this is the case how can we expect our assistant to follow the rules? An acharya shall never go against the laid out procedures and systems.

If you keep your area dirty, others make it dirtier.

5.8   Be a follower first

People normally dream to be a leader, but fortunately or unfortunately, it is imperative to be a follower if one wants to become a Leader. One who can follow can lead. First we need to respect and follow the systems designed and developed either by our predecessors or by ourselves. When people do not follow our instructions or systems suggested we need to understand as to why they are not following the procedures drawn by us, or why they are following a different procedure. One needs to understand that if a system is followed by people for a long time, there must be something good in it. No one would like to follow a system if it is not helping him to grow and achieve his goals. Our predecessors struggled to live successfully in the circumstances they had and made the path easy and safe for their successors and for that they were respected by their followers. They developed the language, letters and numbers, science, logics, thoughts and systems and so on. The logics and systems were developed by the experiences of the people. People were convinced by the Values, Acts and Intentions of elders who followed the good systems that could help the community and respected them. People support the one who respects their feelings and values. If one is not ready to follow the laid procedures, others would not like to follow him. The developments are to be brought gradually by winning the people working with us and not by forcing them or violating the laid procedures.

5.9   Combination of science and ethical values

One who is aspiring to bring a change for betterment of people cannot stop just at this place being a follower. One needs to study the difference in the situation and understand the root cause of the change and then adopt himself to the changes. Just because someone is successful, all his acts need not be a benchmark for us. We should have the logics to decide the acts suitable for our situation and should have the capability of communicating it appropriately to the people by understanding their state of minds. This can be done by working with people, for which we need to follow the systems and sayings that have deep root and analyse the pros and cons while working with it. Just by rejecting a deep rooted system without having ability to convince the masses following it shall make us separate, and we cannot achieve the results we wanted. We should always bear in mind the wordings of late Dr. D. V. Gundappa (a scholar in Kannada) in his book "*MankuThimmanaKagga*" that "A tree is glorious when it has new leaves combined with old deep roots. A combination of new scientific ideas with deep rooted principles, ethics and values is the need for the wellbeing of humans. The latest technology backed by science supported by the values and philosophies of the ancient Sages can bring prosperity to mankind"

The beauty of a tree lies in its fresh leaves coupled with old deep roots. Similarly the combination of new ideas of science with the base of values and ethics explained by old rishis is needed for the survival of mankind –

D.V. Gundappa

Jaya Row of Vedantha Vision, in her speech at 68th All India Textile Conference at Mumbai on 1st Dec 2012 used the term "State of Effortless Excellence" which could be achieved by keeping the worries out of way, foolishness out of way and selfishness out of way.

5.10   Conclusions

Just by adopting latest technology cannot give the required results unless the people working are dedicated and have competency to handle that technology. Having faith in the systems and following it religiously is an imperative of a quality person. The discipline at all time is a basic imperative of success. Keeping worries, foolishness and selfishness out of the way is essential to achieve the state of Effortless Excellence. Respecting the feelings of people around us is very important. One needs to become real acharya. The developments need to address the basic ethics, principles and values that are supporting the development and sustenance of mankind. A combination of new scientific ideas with a backing of deep rooted ethics and values is the need of the day. Self-analysis with the use of five golden questions helps in excelling and becoming a quality person. The quality people indeed are the key to excellence.

Role of management in improving work quality

6.1   Understanding the organization culture

The management plays a very important role by being a model in implementing good systems that can change the attitude of people working for the company. Over the time, the closely bonded or fragmented social systems establish an organization's culture, which reflects conventional behaviour of a company that encompasses beliefs, customs, knowledge, standardization efforts, conflict management and other general work practices. This culture widely influences human behaviour, even though it seldom enters into their conscious thought.

People depend on established cultural systems as they give them stability, security, understanding, and the ability to respond to any given situation. This is why people fear change to their routine practices. They fear that the existing work system will become unstable, their security will be lost, they will not understand the new process, and they will not know how to respond to the new situations. A set of organizational practices, operations, culture, environment and their interaction with products/services exhibit its management system. A good management cautiously works to stabilize good systems where people can be comfortable. A feeling of insecurity, lack of recognition, lack of scope for promotions, no challenge in the job, doing the same work monotonously, etc., demoralises the people.

Employee and group attitude is a key element for measuring the success or failure of an organization. Gradual changes to attitude very likely affect and build a person's work behaviour. Imran Rana observes that like product variation, behavioural variations exist in the shape of conflict which is always there in organizations. This often arises between people or groups in competition for gaining resources, authority, power, attention and status. The competition in implementing their idea is also seen. Organizations generally attempt to engineer social change to their management systems by means of establishing and revising policies, laws, incentives, or coercion to prevent conflicts. Due to these market forces, quality managers are now burdened for one more responsibility of working with HR to stitch off required attitudes in

8.5   How do we compare with our competitors?

In a running race, in order to become a winner, the speed at which I am running is less important than the distance I am keeping ahead of my nearest competitor. Although we achieve the results as anticipated, we cannot be happy for what we achieved, as our success and survival also depends on the competency of our competitors. If our competitors are doing a better job, we are certainly going to lose. Therefore it is always necessary to keep a watch on the competition and develop our systems to achieve better results. We need to benchmark the best systems in each criterion, and work towards meeting and overtaking it. Re- engineering the systems is essential depending on the situation.

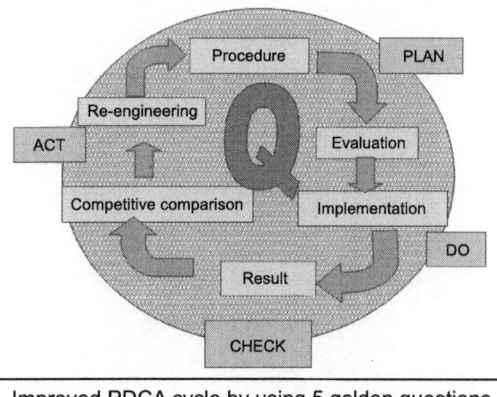

Improved PDCA cycle by using 5 golden questions

The above questions are universal and can be applied to all situations, and in all organizations, for all activities. This is an extension of PDCA concepts developed by Dr. Deming and incorporation of competitive comparison and benchmarking. The procedure and evaluation of procedure is for PLAN, implementation is same as DO in Deming's wheel. Results and competitive comparison is the process of CHECK and Re-engineering is the process of ACT. This can be illustrated as a cycle as shown in figure. The elements are procedure, evaluation, implementation, result, competitive comparison and re-engineering. This requires continuous improvement in the quality of work.

If you are always clear about the activity and the results, you can always emerge as a winner. The above questions are simple and self- explanatory. Anyone can make use of these questions. Here we need to answer ourselves, and there is no need for a third party audit or certifications. If you are satisfied as a winner, what is the use of certificate from others? You should win and you should be happy.

Index